AF587371

MATERIALS SCIENCE TECHNOLOGIES

NEW DEVELOPMENTS IN MATERIALS SCIENCE

MATERIALS SCIENCE TECHNOLOGIES

Additional books in this series can be found on Nova's website
under the Series tab.

Additional E-books in this series can be found on Nova's website
under the E-books tab.

MATERIALS SCIENCE TECHNOLOGIES

NEW DEVELOPMENTS IN MATERIALS SCIENCE

EKATERINE CHIKOIDZE
AND
TAMAR TCHELIDZE
EDITORS

Nova Science Publishers, Inc.
New York

For permission to use material from this book please contact us:
Telephone 631-231-7269; Fax 631-231-8175
Web Site: http://www.novapublishers.com

Library of Congress Cataloging-in-Publication Data

New developments in materials science / editors, Ekaterine Chikoidze and
Tamar Tchelidze.
p. cm.
Includes index.
ISBN 978-1-61668-852-3 (softcover)
1. Materials science. I. Chikoidze, Ekaterine. II. Tchelidze, Tamar.
TA403.N425 2009
620.1'1--dc22

2010013739

Published by Nova Science Publishers, Inc. † New York

Errata for *New developments in materials science*

Editor: Ekaterine Chikoidze and Tamar Tchelidze; **ISBN**: 978-1-61668-852-3

p. iv

The Library of Congress Cataloging-in-Publication Data should be:

New developments in materials science / editors, Ekaterine Chikoidze and Tamar Tchelidze.
p. cm.
Includes index.
ISBN 978-1-61668-852-3 (hardcover)
1. Materials science. I. Chikoidze, Ekaterine. II. Tchelidze, Tamar.
TA403.N425 2009
620.1'1--dc22
2010013739

CONTENTS

PREFACE

Materials science is an interdisciplinary field involving the properties of matter and its applications to various areas of science and engineering. This science investigates the relationship between the structure of materials at atomic or molecular scales and their macroscopic properties. It includes elements of applied physics and chemistry. With significant media attention focused on nanoscience and nanotechnology in recent years, materials science has been propelled to the forefront at many universities. This new book reviews research and presents new developments in the field of materials science.

Chapter 1 - Changes in the quantum well properties caused by periodic ridges in the surface were studied within the limit of quantum theory of free electrons. The authors show that due to destructive interference of de Broglie waves, some quantum states become quantum mechanically forbidden for free electrons. Wave-vector density in k space is reduced dramatically.

At the same time the number of free electrons does not change considerably, as the metal remains electrically neutral. Because of the Pauli exclusion principle, some free electrons must occupy quantum states with higher wave numbers. The Fermi vector and Fermi energy of quantum well increase, and consequently, the work function decreases. Resent experiments are in qualitative agreement with the presented theory.

This effect could exist in any quantum system comprising fermions inside a potential-energy box of special geometry.

Chapter 2 - It was investigated the samples of Bi-Pb-Sr-Ca-Cu-O system, fabricated by different methods: solid state reaction and melt quenching ones under the influence of concentrated solar radiation.

It was studied the superconducting properties of samples by the supersensitive mechanical method, where for a sample suspended by a thin elastic thread and performing axial- torsional oscillation in a transverse magnetic field, it was recorded the temperature dependence of oscillations dissipation and period. The authors also used standard methods: magnetic moment, electrical resistance and magnetic susceptibility measurements.

Chapter 3 - In the presented paper optical and paramagnetic spectra, besides that thermally stimulated luminescence (TSL) were investigated for LiF(OH) crystals irradiated at low (20 K and 100K) and at usual nuclear-pile temperatures. Some of the samples were irradiated in free or compressed states. It was established the dose dependence of the behavior of hydrogen defects formed after decay of hydroxyl ions under the action of ionizing radiation.

Chapter 4 - The effect of samarium ion valence on the optical properties of samarium diantimonide thin films has been studied. The optical constant spectra were obtained by processing of the reflectance spectra of $SmSb_2$ films at different values of samarium ion valence (+2.7; +2.4; +2.2) by the Kramers-Croning method. At all the values of samarium ion valence, there remained a gap in the energy spectrum and a clear plasma edge in the spectral dependence of reflection coefficient. The decrease in valence causes the decrease in plasma frequency and in the effective number of charge carriers contribution to plasma oscillation and also reduces the role of f-electrons in the formation of pf and fd hybrid states. The energy zone extremum is very sensitive to changes in the valent state of samarium ion as well.

Chapter 5 – The authors present the first systematic study of timing and frequency diagrams of magnetic video-pulse influence on the NMR two-pulse echo in a number of magnets (ferromagnets, ferrites, half metals, manganites). It is shown that the dependence of two-pulse echo intensity on the temporal location of a magnetic video-pulse in respect to radio-frequency pulses and spectral diagram of this influence are defined mainly by local hyperfine field anisotropy and domain walls mobility. These diagrams could be used for the identification of the nature of NMR spectra in multidomain magnetic materials and to improve the resolution capacity of the NMR method in magnets.

Chapter 6 - It is shown experimentally and proved theoretically, that the reason of low temperature melting of a crystal at pulse laser annealing is the electronic mechanism of attenuation and isotropization of chemical bonds between its atoms. The formula is received for critical concentration of anti-bonding quasi-particles - electrons in conduction band and holes in valence band, above which a low temperature melting of the semiconductor on the electronic mechanism should be occurred. Possible mechanisms of non-thermal melting of semiconductors on an example of crystals with a covalent bond are discussed.

Chapter 7 - A model of a divacancy which explains a number of experimental facts in irradiated Ge and GaAs is proposed. The model explains the absence of photoconductivity associated with photoexcitation of infrared absorption bands at 0.44, 0.52 eV (Ge) and 0.80, 1.00 eV (GaAs) respective to divacancies; the difference between the values of activation energy of divacancy energy levels are determined by different methods; the existence of the limiting position of the Fermi level in the forbidden gap in the case of irradiation with large radiation fluxes is revealed.

Chapter 8 - Variety of boron crystalline modifications and boron-rich solids are known to be constructed from the interconnected boron icosahedra. The same short-range structural order is characteristic for amorphous boron. Then, most of B atoms usually are members of the almost regular atomic triangles. This circumstance leads to possibility of the new kind of elemental boron nanostructures in form of flat or rolled boron surfaces with triangular two-dimensional lattices.

Based on the given structural description and using appropriate B−B interatomic potential it is possible to calculate easily their ground-state parameters, as well as the molar binding energy, equilibrium bond length, zero-point vibration frequency etc.

Thereto the quasi-classical B−B potential in dependence on interatomic distance suggested earlier seems to be useful because part of its parameters, like the coefficients of harmonic and anharmonic terms (when the potential-function in the vicinity of equilibrium interatomic distance is approximated by the cubic expression) and atomic core charge (which is near the boron atom valence of 3) had been successfully applied to interpret isotopic composition effect on boron structural and melting parameters, respectively.

There are estimated set of quasi-classical ground-state parameters for the stable boron nanotubes. In particular, corresponding molar binding energy is found to be ~ 8.40 eV / atom.

Chapter 9 – The authors present the measurements and analytical expressions for the magnetization at T=77 K of the high-temperature ceramic superconductor $YBa_2 Cu_3 O_{7-x}$. The behaviour of magnetization is studied as a function of external magnetic field. The sample was prepared by the method of solid-state reaction. Magnetization measurements were conducted with ballistic method. Analytical expressions are obtained in Bean's critical state model by taking into account the field H_{c1}. Description of experimental hysteresis of the polycrystal structure of the sample and granule's anisotropy are taken into account as well.

Chapter 10 - Carbon is usually a foreign unforeseen impurity in semiconductor materials in contrast to elements of III and V group of Periodic System. But carbon is also widely used in construction of modern devices. Therefore, investigation of carbon behavior, as an impurity, especially, in SiGe alloys is very relevant. Si-Ge alloys attract considerable interest in advanced thermoelectric, electronic and optoelectronic applications. There are few data in literature under this question. Si and SiGe alloys crystals have been grown by the Czochralski method. By means of IR spectroscopy at 300 and 77K temperatures we have determined the carbon content and its distribution in Si and SiGe alloys crystals. Obtained regularity of carbon distribution in SiGe alloys should be explained by high activity of carbon relatively to Si and slight activity relatively to Ge.

Chapter 11 - A process and a mechanism of stimulation of a low-temperature plasma anodization using ultraviolet radiation in the case of the formation of oxide films of GaAs is proposed. Insulation of the active elements of integrated circuits based on intrinsic GaAs oxide was performed, which provides a significant decrease of leak currents, increases the thermal stability and breakdown field. The stimulating effect of ultraviolet radiation on the process of plasma anodization is attributed by the appearance of additional concentration of antibonding quasiparticles (electrons and holes), which weakens the chemical bonds in the GaAs.

Chapter 12 - Thin ferromagnetic films (Fe, Ni) on GaAs substrates have involved as a model system for the integration of magnetic materials with semiconductors. For practical applications, it is highly desirable that the injection of spin currents should be electrical and the injecting from a classical ferromagnetic metal in a metal/semiconductor heterostructure is most direct way for spin injection. In majority of investigations metallic thin films has been deposited at elevated temperatures, however, due to the diffusion of Ga and As into the film even at room temperature no sharp interface, satisfying requirements of spintronic devices, is formed. In order to reduce intermixing effects, Fe was recently deposited also at room temperature, and good epitaxial growth was reported without formation of a dead magnetic layer. Here we report the physical properties of structure obtained via electrochemical deposition metallic (Fe, Ni, Pd) thin films on GaAs substrate. Electrochemical deposition is a low-energy process and offers inexpensive alternative for potentially fabricating abrupt Ferromagnetic Metal/GaAs interfaces. Thin ferromagnetic films on GaAs substrates were obtained by electrochemically deposition of ferromagnetic metal on the previously electrochemically etched surface of semiconductor. The aqueous solution of chloride was used for deposition of metals. There were studied current-voltage and capacitance-voltage characteristics for determination interfaces quality, and the results of investigation physical properties of interface are presented in this paper.

Chapter 13 - The technology of receiving of technical Si or/and magnesium silicide has been described. Presented technology is based on the magnesium thermal restoration method. In the technology offered by us there has been applied low-quality, cheap, local quartzite in which content of $Si0_2$ does not exceed 95% and therefore this quartzite can not be applied without its ore-dressing. The process is strongly exothermal and in the case of selection of suitable conditions the process goes on at the expense of own heat. According to the quantity of added magnesium it is possible to obtain Si, magnesium silicide or both of them. The expenditure of energy is reduced marginally.

We studied the influence ratio of quartzite/ Mg, fractional composition of quartzite, duration of the process and initial temperature at the restoration process. Values of the noted factors, which provide achievement of maximum degree of technical Si and magnesium silicide purity during the production, have been established.

Chapter 14 - In the present paper, using the discontinuous Fe films as examples the authors have investigated the influence of the shape of magnetic particles on the magneto-optical and optical properties of the ultrafine structures. Obtained results have confirmed the significant change of the components of the tensor of effective dielectric permittivity and subsequently of the magneto-optical and optical properties which was brought about by the change of the shape of magnetic particles.

These calculations proved that if the authors take into account the shape of the particles, they will achieve a perfect match of the experimental results and the theoretical calculations.

Chapter 15 - The energetic conditions essential for graphite transformation into the diamond are formed by using the energy of explosion and neutron irradiation.

First the initial metal-carbon system is subjected to loading by weak shock waves (pressure $\leq$ 1,75 GPa). The graphite is transformed into the activated state with a rhombohedric structure. Further the irradiation with fast neutrons under low-temperature conditions (300K) is performed. The neutron dose are about $3 \cdot 10^{22}$ N/m^2; energy –E $\geq$0,1mev.

As a result of such complex treatment, the phase transformation of graphite into the diamond is accomplished.

The neutron irradiation is the key stage in the proposed technology. Available reactors can be used for implementation of this stage.

Chapter 16 - In the present paper, using magnetite magnetic fluids as examples, we consider theoretically and experimentally the optical and magneto-optical properties of magnetic fluids based on particles of magnetic oxides, for which the relation $k^2 << n^2$ holds (n and k are the optical constants of the material).

In this work the equatorial Kerr effect is represented within the theoretical Maxwell-Garnett model. A theoretical analysis has shown that in the mentioned case of magnetic oxides, the equatorial Kerr effect $\delta_e(q)$ for magnetic fluids with the ratio of the volume occupied by magnetic particles is related to the equatorial Kerr effect δ_m on the material of particles by the simple relation.

The estimations showed that the amount of effect is proportional to the occupancy of the volume of the magnetic fluid with magnetic particles - q. Besides, the spectral dependences of magneto-optical effects are similar for the magnetic fluids and for the magnetic particles. The experimental and theoretical results for the magnetite magnetic fluids are in a good agreement.

The result is expected to be applicable to all the magnetic ultrafine medium, the optical constants of which, n and k, holds this relation $k^2 << n^2$.

Chapter 17 - Scientific investigations show that hydrolysis rate of native cellulosic compounds under enzymatic fermentation is low and therefore pretreatment of cellulosic wastes becomes one of the topical points. Cellulose pretreatment method based on combination of chilling and mechanical milling of cellulosic wastes was used for investigation of structural changes of different cellulosic materials. When frozen water contained in cellulosic wastes expands it causes the mechanical stress that is followed by partial destruction of crystalline cellulose. Cellulose decomposition rate, changes of both: cellulose polymerization degree and specific surface area have been determined when cellulosic wastes such as: sawdust, cardboard, newsprint, filter paper and composite of all above materials were frozen at 0, -20, -70 and -196 ^{0}C. Studies showed that mechanical milling of frozen cellulosic wastes enhances subsequent breaking of cellulose that leads to further increase of cellulose specific reactionary surface accessible for enzymes. Investigations conducted under the proposed research showed that maximum increase of specific surface area of cellulosic wastes could be reached when cellulosic wastes where chilled at -20^{0}C and milled at once.

Chapter 18 - Deposition of SiO_2 films was made from the solution prepared on the basis of tetraethoxysilane. The “maturing” process of film-forming solution was controlled by high-resolution nuclear magnetic resonance (nmr) method (on hydrogen nuclei) and, at the same time, was checked by testing depositions. The deposition of films was carried out by the method of centrifugation on “кдб-10”- grade silicon plates having the hole conductivity and <111> crystallographic orientation. For determination of the composition of films we studied infra-red (IR) spectra registered on “UR-20” spectrometer.

On the basis of the obtained films, the metal-dielectric-semiconductor Al-SiO_2-Si-Al structures were prepared. Some electro-physical characteristics have been studied: electric strength, dielectric permeability and dielectric losses. The measurement of these characteristics revealed the prospects of the obtained films both, in microelectronics and in nanotechnologies.

Chapter 19 - The last 10-15 years have been characterized by intensive research & develop on renewable energies including photovoltaic (PV) energy conversion. In this context the concentrator technology was researched more intensively. One reason is the availability of high-efficient A3B5 based triple-junction solar cells which have been originally developed for space applications. Applying now the concentrator technology an application on earth is possible. The significantly higher cost of the A3B5 technology can be compensated on system level and can be even lower compared to the standard silicon PV approach.

Chapter 20 - The character of thermal recovery of mechanical properties (yield stress, plasticity, microhardness) of LiF crystals irradiated in the mixed (n, γ) field of reactor at 20 and 300 K temperatures has been studied. The evolution of radiation-induced microstructure is tracked, and it is shown that alongside with the disappearance of elemental defects and their complexes, the new structures – the prismatic dislocation loops of interstitial type and microcavities (pores) are formed.

In: New Developments in Material Science ISBN 978-1-61668-852-3
Editors: E. Chikoidze and T. Tchelidze

Chapter 1

QUANTUM STATE DEPRESSION IN A QUANTUM WELL OF SPECIAL GEOMETRY

Avto Tavkhelidze**, *Vasiko Svanidze and Irakli Noselidze
Tbilisi State University,
Chavchavadze Avenue 13, 0179 Tbilisi, Georgia

ABSTRACT

Changes in the quantum well properties caused by periodic ridges in the surface were studied within the limit of quantum theory of free electrons. The authors show that due to destructive interference of de Broglie waves, some quantum states become quantum mechanically forbidden for free electrons. Wave-vector density in k space is reduced dramatically.

At the same time the number of free electrons does not change considerably, as the metal remains electrically neutral. Because of the Pauli exclusion principle, some free electrons must occupy quantum states with higher wave numbers. The Fermi vector and Fermi energy of quantum well increase, and consequently, the work function decreases. Resent experiments are in qualitative agreement with the presented theory.

This effect could exist in any quantum system comprising fermions inside a potential-energy box of special geometry.

* E-mail: avtotav@gmail.com

1. INTRODUCTION

Recent developments of nanoelectronics enable the fabrication of structures with dimensions comparable to the de Broglie wavelength of a free electron inside a solid. This new technical capability makes it possible to fabricate some microelectronic devices such as resonant tunneling diodes and transistors, superlattices, quantum wells (QW), and others [1] based on the wave properties of the electrons. In this article, we discuss what happens when regular ridges, which cause interference of de Broglie waves, are fabricated on the surface of a QW layer. We will study the free electrons inside a rectangular potential-energy box with ridged wall and compare the results to the case of electrons in a box with plane walls. We have shown that modifying the wall of a rectangular potential-energy box leads to an increase of the Fermi energy level. Results obtained for the potential-energy box were extrapolated to the case of QW layers.

2. ELECTRONS IN A POTENTIAL-ENERGY BOX WITH A RIDGED WALL

We begin with the general case of electrons inside a potential-energy box. Assume a rectangular potential-energy box with one of the walls modified as shown in Figure 1. Let the potential energy of the electron inside the box volume be equal to zero, and that outside the box volume be equal to infinity. The ridges on the wall have the shape of strips having depth a and width w. Let us name the box shown on Figure 1 as ridge potential-energy box RPEB to distinguish it from the ordinary potential-energy box PEB having plane walls. The time independent Schrödinger equation for electron wave function inside the PEB has the form

$$\nabla^2\Psi + (2m/\hbar^2)E\Psi = 0 \tag{1}$$

Here, Ψ is the wave function of the electron, m is the mass of the electron, and E is the energy of the electron. Let us rewrite Eq. (1) in the form of the Helmholtz equation,

$$(\nabla^2 + k^2)\Psi = 0 \tag{2}$$

where k is wave vector, $k=\sqrt{2mE}/\hbar$. Once the ridge depth a in our particular case is supposed to be much less than the thickness of the metallic film $a<<L_x$, we can use the volume perturbation method to solve the Helmholtz equation [2].

The idea is as follows: The whole volume is divided in two parts, the main volume MV and the additional volume AV. The MV is supposed to be much larger than the AV, and it defines the form of solutions for the whole composite volume. Next, solutions of the composite volume are searched in the form of solutions of the MV. The method is especially effective in the case where the MV has a simple geometry, for example, a rectangular geometry, allowing separation of the variables. In our case, the whole volume in Figure 1 can be divided in two, as shown in Figure 2. We regard the big rectangular box as the MV and the total volume of strips as the AV. The MV has dimensions L_x, L_y, and L_z. The solutions of Eq. (2) for such a volume are well known. Because of the rectangular shape, solutions are found by using the method of separation of variables. The solutions are plane waves having a discrete spectrum,

$$\mathrm{k}_n^{mx}=\pi n/L_x,\ \mathrm{k}_j^{my}=\pi j/L_y,\ \mathrm{k}_i^{mz}=\pi i/L_z \tag{3}$$

Here, k^{mx}, k^{my} and k^{mz} are the x, y, and z components of wave vectors of the MV and n,m, i=1,2,3,... In the same manner the spectrum of the single strip of the AV is the following:

$$\mathrm{k}_p^{ax}=\pi p/a,\ \mathrm{k}_q^{ay}=\pi q/w,\ \mathrm{k}_i^{az}=\pi i/L_z \tag{4}$$

k^{ax}, k^{ay} and k^{az} are components of wave vectors of one strip of AV, p,q=1,2,3,..., and a and w are dimensions of the strip, as shown on Figure 2. Analysis conducted in [3] shows that wave vector spectrum of RPEB is

$$\mathrm{k}^{\mathrm{x}}{}_p=\pi p/a,\ \mathrm{k}^{\mathrm{y}}{}_q=\pi q/w,\ \mathrm{k}^{\mathrm{z}}{}_i=\pi i/L_Z \tag{5}$$

and is equal to the spectrum of a single strip Eq. (4). In the Eq. (5), we skip some working indices used in this section to simplify the presentation.

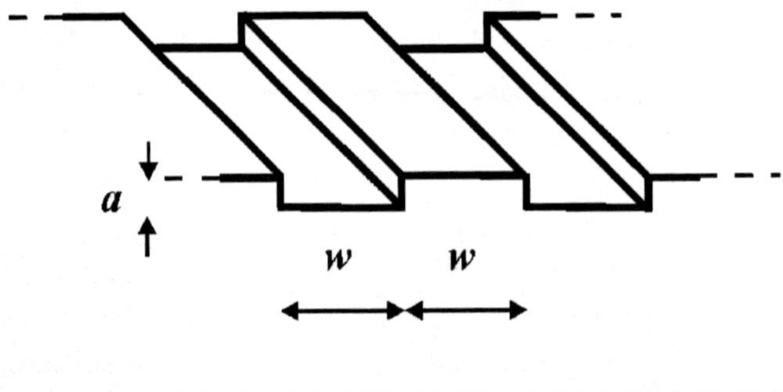

Figure 1. 3D view of ridged potential-energy box.

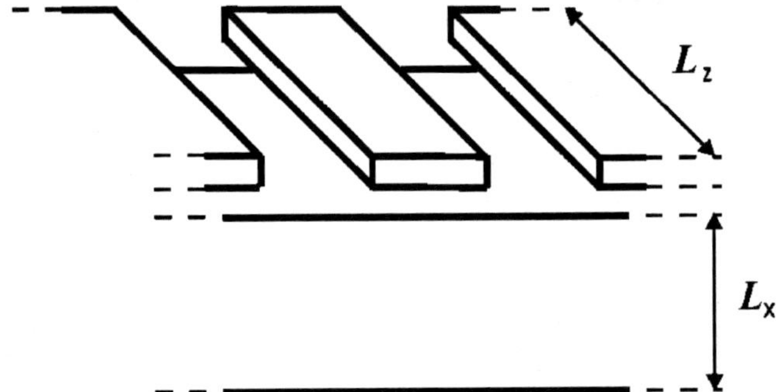

Figure 2. Potential-energy box with ridged wall divided into two volumes.

Equation 5 is obtained using the volume perturbation method for solving of the Helmholtz equation. This method assumes that the AV is much less than the MV $(a/2L_x) \ll 1$. Special attention should be paid to the limit of very low a and w. The case $a, w \to 0$ has the following physical interpretation. Standing waves in the MV ignore the AV because of wave diffraction on it. If we assume that the wave ignores nonregularities with dimensions less than its wavelength, we must make the following corrections in Eq. (5). It is valid for $k^x > 2\pi/a$ and $k^y > 2\pi/w$ for p, q=2, 3, 4.... For the range $0<k^x<2\pi/a$ and $0<k^y<2\pi/w$, Eq. (3) should be used for RPEB instead of Eq. (5). In practice, dimensions a, w are such that only the first few k should be added to Eq. (5).

For the case of large $a/2L_x$, other methods were used. The general solution of Eq. (2) in such a complicated geometry exhibits several problems. A complicated surface shape does not allow to find an orthogonal coordinate system that will allow to separate the variables. Therefore, boundary conditions may be written only in the form of piecewise regular functions. A general solution of Eq. (2) usually contains infinite sums. However, there are methods [2] that allow one to obtain a dispersion equation and to calculate the wave vector. The Helmholtz equation is frequently used for calculating the electromagnetic field in electromagnetic resonator cavities, waveguides, and delay lines. We found the following similarities between the electron wave-function inside the RPEB and electromagnetic field inside electromagnetic delay lines. First, our geometry matches the geometry of a corrugated waveguide delay line [4] . Second, the same Eq. (2) is used to describe both cases. Third, the boundary condition for the electromagnetic wave inside the corrugated waveguide $\epsilon=0$ (here, ϵ is the electric component of the electromagnetic wave) for walls of a conductive waveguide exactly matches the boundary condition for electrons, $\Psi=0$ outside the metal. Fourth, any wave can be presented as the sum of plane waves in both cases. We also found that similar analogies are described in the literature [5]. Therefore, for the case of high a, we used the method of solving the Helmholtz equation inside the corrugated waveguides. This method is based on solving a transcendental equation. We found numerical solutions for the transcendental equation for high $a/2L_x$ and obtained the same result, namely, the reduction of the spectrum density for RPEB relative to PEB.

For very thin layers $L_z >> L_x$, w, it is expected that our structure will not exhibit quantum features in the Z direction. Therefore, it is reasonable to consider a model of the electron motion in a two-dimensional (2D) region delimited by the line X=0 from one side, and a periodic curve on the opposite side. In that case, we can consider the 2D Helmholtz equation and use special methods to solve. Specifically, we used the Boundary Integral Method (BIM), which is especially effective for the low-energetic part of the spectrum and has been widely employed for studying 2D nano systems [6].

The BIM method implies consideration of an appropriate integral equation instead of the Helmholtz equation. We have applied the corresponding numerical algorithm [7] to a finite number of periods and calculated the lowest few ten energy levels. We note that computation of higher energy levels with reasonable accuracy demands rapidly increasing machine-time. The obtained spectrum of transverse wave vectors also shows a reduction in spectral density.

At the end of this section, it is worth to note that the above-discussed effect of the energy-spectrum reduction is closely related to the so-called Quantum Billiard Problem. The 2D system studied represents a modification of the quantum billiard problem. Unlike the circle and rectangular billiards, the corrugated boundary makes the system under consideration non-integrable. This means that the number of degrees of freedom exceeds the number of constants of motion. Such billiards constitute chaotic systems. The distinction between the spectra of chaotic and non-chaotic (regular) systems is exhibited by an energy level spacing distribution [7, 8]. Namely, consecutive energy levels are likely to attract each other in the case of an integrable system, whereas they repel each other in the case of a chaotic system.

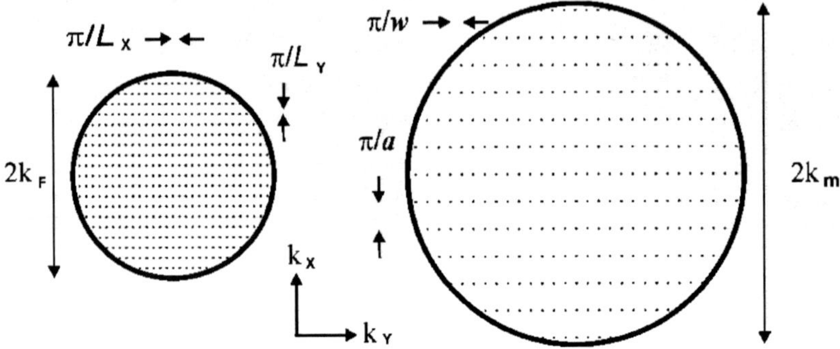

Figure 3. Section of Fermi sphere is *k* space for PEB (left) and RPEB (right). Points denote quantum states for free electron.

3. Free Electrons in Metal Layer with Ridged Surface

To investigate how periodic ridges change quantum states inside the thin metal layer, we will use a quantum model of free electrons. Free electrons inside the metal form a Fermi gas. Cyclic boundary conditions of Born-Carman,

$$k^x{}_n = 2\pi n/L_x,\ k^y{}_j = 2\pi j/L_y,\ k^z{}_i = 2\pi i/L_z, \qquad (6)$$

are used instead of Eq. (3). Here, n, j, i=0, ±1, ±2, ±3... The result of the theory is a Fermi sphere in k space. All possible quantum states are occupied below k_F at T=0. However, for T>0, there are two types of free electrons inside the Fermi gas. Electrons with $k \approx k_F$ interact with their environment and define the transport properties of metals such as charge and heat transport. Electrons with $k \ll k_F$ do not interact with the environment because all quantum states nearby are already occupied by other electrons (it becomes forbidden to exchange small amounts of energy with the environment).

Such electrons are ballistic and have formally infinite mean free path. This feature allows us to regard them as planar waves, traveling between the walls of the metal (if the distance between walls is not too great). Further, we will concentrate on such ballistic electrons.

Once we work with electrons with infinite (or very long) mean free paths, we can regard the metal layer as a potential-energy box and extrapolate calculations of the previous section to it.

We start by comparing the volume of an elementary cell in k space for thin metal films with and without periodic ridges. From Eq. (5) and Eq. (6), we have:

$$dV_k=8\pi^3/(L_xL_y\,L_z) \text{ and } dV_{km}=8\pi^3/(awL_z) \quad (7)$$

Here, dV_k is the volume of an elementary cell in k space for a plain film, and dV_{kin} is the volume of an elementary cell in k space for an ridged film. Further, we find maximum wave vector and Fermi energy as in reference [3]

$$k_m= k_F\,[L_y(L_x+a/2)/(aw)]^{1/3}. \quad (8)$$

$$E_m=E_F[L_y(L_x+a/2)/(aw)]^{2/3} \quad (9)$$

If we assume $a<<L_x$, Eq. (9) can be rewritten in the following simple form:

$$E_m=E_F(L_xL_y\,/aw)^{2/3}. \quad (10)$$

The dimensional quantum effects in ultra-thin metal films, using a rectangular potential-box model and quantum model of free electrons, were studied in Ref. [9]. Quantum state depression in metal films was investigated experimentally in Ref. [10]. Quantum state depression in semiconductor ridged quantum well was investigated theoretically in Ref. [11].

CONCLUSIONS

To investigate a new quantum interference effect in metal layers, we studied the behavior of free electrons in the potential-energy box of special geometry. It was shown that when periodic ridges are introduced in the plain wall of a rectangular potential-energy box, the spectrum density of possible solutions of the Schrödinger equation is reduced dramatically. Once the number of possible quantum states decreases, electrons must occupy higher energy levels because of the Pauli exclusion principle. Results obtained for electrons in the modified potential-energy box were extrapolated to the case of the free ballistic electrons inside the metal film. The electron distribution function changes. Fermi energy increases, and consequently, work function decreases. Recent experiments demonstrate quantitative agreement with the theory. Increase in the Fermi level and the corresponding decrease of the work function of the thin films will have practical use for devices working on the basis of electron emission and electron tunneling. In addition, such layers will be useful in the semiconductor industry, particularly for the structures in which contact potential difference between two layers plays an important role.

REFERENCES

[1] H. Davies, *The physics of low-dimensional semiconductors*, (Cambridge university press, 1998).

[2] V. M. Sedykh, *Waveguides With Cross Section of Complicated Shape* (Kharkhov University Press, Kharkov, 1979), p. 16.

[3] A. Tavkhelidze, V. Svanidze, and I. Noselidze 2007 J. Vac. Sci. Technol. B 25, 1270.

[4] R. S. Elliott, 1954 IRE Trans. *Antennas Propag*. 2, 71.

[5] R. A. Silin and B. P. Sazonov, *Delaying Systems* (Soviet Radio, 1966), p. 395.

[6] G. Chen and J. Zhou, *Boundary Element Methods* (Academic, San Diego, 1992).

[7] I. Kosztin and K. Schulten, 1997 *Int. J. Mod. Phys*. C 8, 293.

[8] F. Haake, *Quantum Signatures of Chaos* (Springer, Berlin, 1991).

[9] V. V. Pogosov, V. P. Kurbatsky, and E. V. Vasyutin, 2005 *Phys. Rev*. B 71, 195410 .

[10] A. Tavkhelidze, A. Bibilashvili, L. Jangidze, A. Shimkunas, P. Mauger, G.F. Rempfer, L. Almaraz, T. Dixon, M.E. Kordesch, N. Katan, and H. Walitzki 2006 *J. Vac. Sci. Technol*. B, 24, 1413.

[11] A. Tavkhelidze, V. Svanidze 2008 *Int. J. Nanoscience*, Vol. 7, No. 6, pp. 333-338.

In: New Developments in Material Science ISBN 978-1-61668-852-3
Editors: E. Chikoidze and T. Tchelidze

Chapter 2

INFLUENCE OF THE TECHNOLOGY OF SYNTHESIS OF THE HTSC SYSTEM BI-PB-SR-CA-CU-O ON POSSIBLE CHANGE OF THE CRITICAL TEMPERATURE T_C OF SUPERCONDUCTING TRANSITION

J. G. Chigvinadze*[*1], D. D. Gulamova[2], S. M. Ashimov[1], T. V. Machaidze[1], O. V. Maghradze[1], G. J. Donadze[3] and D. E. Uskenbaev[2]

[1]E.Andronikashvili Institute of Physics, 0177 Tbilisi, Georgia
[2]The Institute of Materials Science SPA “Physics-Sun” of Academy of Science, Republic of Uzbekistan, 700084, Tashkent, Uzbekistan,
[3]Georgian Technical University, Tbilisi, Georgia.

It was investigated the samples of Bi-Pb-Sr-Ca-Cu-O system, fabricated by different methods: solid state reaction and melt quenching ones under the influence of concentrated solar radiation.

It was studied the superconducting properties of samples by the supersensitive mechanical method, where for a sample suspended by a thin elastic thread and performing axial- torsional oscillation in a transverse magnetic field, it was recorded the temperature dependence of oscillations

[*] E-mail: chigvinadze@yahoo.com, jaba@iphac.ge

dissipation and period. We used also standard methods: magnetic moment, electrical resistance and magnetic susceptibility measurements.

Above the critical temperature of the main phase Bi(2223) superconductive transition with T_c=107,5K was observed a "chaos" region in the 118-135 K temperature interval where it was fixed some separate bursts of frequency and dissipation of oscillations.

It is supposed that the "chaos" region could point to the possibility of existence of other magnetic or more high-temperature phases in small concentrations as single separate islands in the normal material of 2223 – phase.

The creation of high-temperature oxide superconductors with a high critical characteristic is an important problem both from the point of view of the fundamental problems of HTSC and the solving of practical problems. The widely known solid state reaction method does not provide a high-density critical current as result of granular structure formation which is the reason of weak Josephson links formation on intergranular boundaries. The melt technologies could eliminate the shortages of the solid-state reaction method [1,2].

The use in the glass-crystal technologies a concentrated solar energy as a source of heating has some advantages as compared with a widely applied synthesis method based on the solid state reactions. For example, such type of heating excludes the introduction of impurities from technology equipment and the melt synthesis is realized in the ozone medium, created at air ionization under influence of the concentrated solar radiation, increasing the Cu^{2+} cations amount.

If the process of preparation of 2201 and 2212 phases on basis Bi-compounds is relatively simple, a composition 2223 phase with a higher value of critical temperature T_c is very sensitive to outer conditions. In some articles [3-6] it is pointed to facts, that if one increases Ca and Cu concentration, it is possible to get 2234 ore 2245 phases with their T_c higher, than for 2223.

To find out peculiarities of the formation of superconductive phases fabricated using a concentrated solar radiation it was synthesized and studied nominal 2223 composition samples. Besides the aim of synthesis and study of the 2223 composition superconductive phase properties, it was supposed to find out and study the superconductive properties of phase having possibly higher critical temperature of transition into the superconductive state T_c. In this way it should be revealed the difference between properties of similar nominal composition samples but fabricated by different methods: the solid state reactions method and the method of melt under influence of the concentrated solar radiation and superfast quenching.

The defined composition mixtures were prepared by the standard ceramic technology and pressed in briquettes. The melting of briquettes was realized under the influence of concentrated solar radiation. The quenching of melt with aim to prepare of amorphous precursors was made by the powderization. The thermal treatment of the amorphous precursors was carried out in the resistance furnaces in the air medium at the temperature range 800-870°C.

The samples prepared by the fast quenching of melt were lumps-plates and nails-viskers. The superconductive phase 2223 with T_c=107,5K was prepared after the thermal treatment of the glass like precursors at 850°C during 60 hours. Its content in a starting nominal composition $Bi_{1,7}Pb_{0,3}Sr_2Ca_2Cu_3O_{10-\delta}$ sample was about 95%.

We investigated all specimens by X-ray method for study their structure.

All samples were thermally treated in similar conditions. It should be noted that Bi(2223) HTSC samples, synthesized by the solid state reaction method, are practically single-phase containing only the traces of the 2212 phase. The sample on the basis precursors, prepared by the superfast melt quenching, contains about 95% of the 2223 phase and impurity phases. In these samples the concentration of phases different from the 2223 phase is small and their common concentration is not more then ~5%.

We used a number of methods for investigation of the superconducting properties of our specimens: supersensitive mechanical method for study pinning and dissipative processes in the installation with superconducting specimens, performing axial-torsional oscillations in a steady magnetic field [7,8], and standard methods for magnetic moment, electrical resistance and magnetic susceptibility measurements.

After superconducting transition in magnetic field in HTSC the Abrikosov vortices are created and during this process the pinned vortices cause the change the period of oscillations of a superconductive cylinder, and vortices torn off the pinning centers stipulate the change of dissipation of the oscillations. So, the study of the temperature dependence of frequency and dissipation gives one possibility to study superconducting properties and allows one to reveal new superconducting phases with higher critical temperatures T_c, if they are in the samples.

When we use mechanical method, we bring the specimen in a certain magnetic field, cool them down to 77K in this magnetic field, slowly increase temperature and start measurements of the temperature dependence of dissipation and period of oscillations.

When temperature is increased from 77K, then the pinning force decreases and the oscillation period increases, and at $T>T_c$ we have constant value of oscillation period. At the same time the damping increases, reaching a maximum, and then decreases to a constant value of δ in the normal state, without any peculiarities at $T>T_c$ (Figure 1).

The standard methods: the magnetic moment, electrical resistance and magnetic susceptibility measurements revealed the same value of T_c=107,5K. These methods do not fix any peculiarities at $T>T_c$ in these specimens.

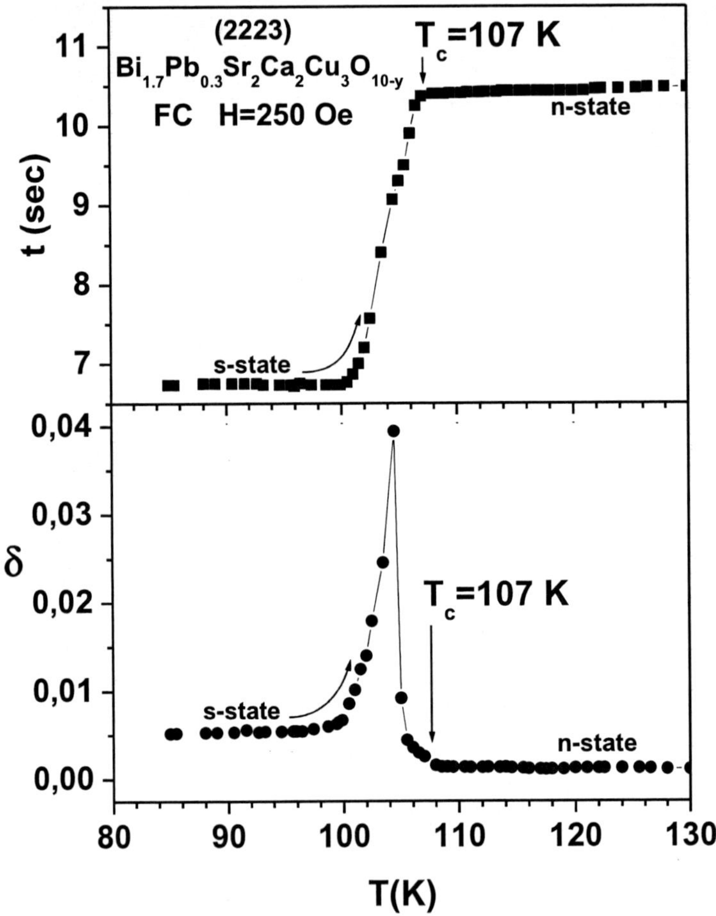

Figure 1. Temperature dependence of logarithmic decrement of damping δ and period t of oscillations in the magnetic field H=250 Oe for a single-phase $Bi_{1,7}Pb_{0,3}Sr_2Ca_2Cu_3O_{10-\delta}$, fabricated by solid state phase technology.

This points to the fact, that in single-phase specimens above T_c it is not observed other more high-temperature superconductive phases.

For samples , fabricated by the superfast melt quenching, on the temperature dependence of the oscillation period t (Figure 2) there are sharply expressed peculiarities not only near T_c of the main HTSC phase (Bi2223) with critical temperature T_c~107,5 K, but also below and higher then temperature 107,5 K (Figure 2).

This apparently points to the fact that in this sample it is present other phases both more low-temperature with T_c of the order of 80-83 K, 90K, 95 K (this apparently is the 2212 phase), and also more high-temperature, new, unknown nature phases (Figure 2), besides the superconducting phase (2223) with T_c~107,5K.

In the temperature range T~118-140K it is clearly seen bursts of period of sample oscillation and this temperature interval is called by us as "chaos" region. It should be noted, that with the increase of the magnetic field the maximum temperature of "chaos" region is displaced to the higher temperature side.

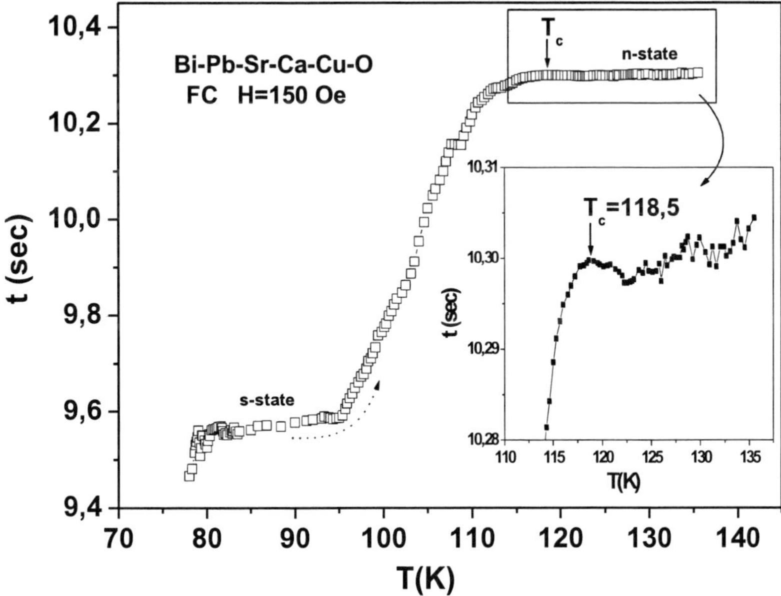

Figure 2. Temperature dependence of period t in the magnetic field H=150 Oe for a $Bi_{1,7}Pb_{0,3}Sr_2Ca_2Cu_3O_{10-\delta}$ sample on the basis of precursors fabricated by the superfast melt quenching.

These results were received on different samples of nominal composition (2223) of Bi-Pb-Sr-Ca-Cu-O system, synthesized by the superfast melt quenching using solar energy. The presented results suppose possibility of presence of traces of superconductive regions up to T~135 K.

The insertion in Figure 2 points to the fact that in this temperature range of the order of 118-119 K it starts a sharp decrease of period of sample oscillations what we relate with (in analogy with Figure 1) critical temperatures T_c. The results of investigations of temperature dependence of the electric resistance of these samples, presented in Figure 3, showed T_c=119 K what is in a good agreement with value T_c defined by mechanical method (Figure 2).

This last circumstance points to the fact that in the investigated Bi-Pb-Sr-Ca-Cu-O samples, synthesized by the superfast melt quenching technology using the solar energy, the high-temperature superconducting phase with T_c≈119 K has a such concentration in the matrix of main 2223 phase that one could manage to record it not only by the supersensitive mechanical method, but also by the standard method of the electrical conductivity measurements.

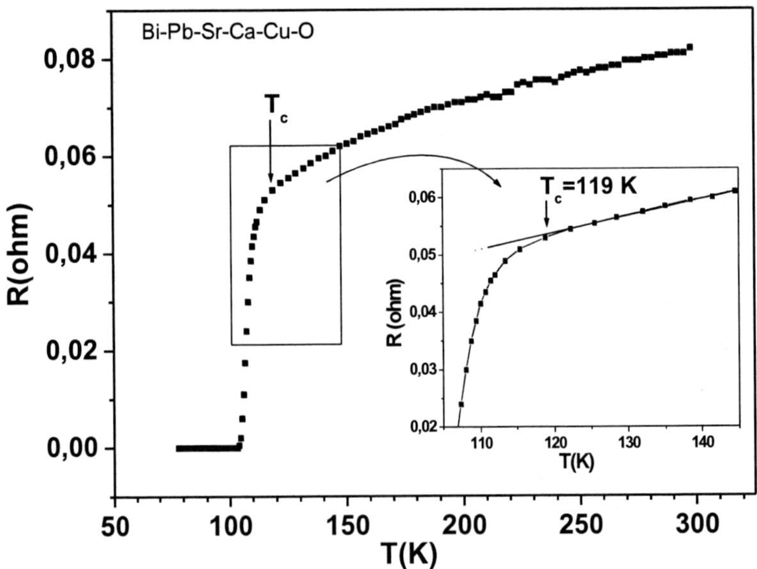

Figure 3. Temperature dependence of electrical resistivity for a $Bi_{1,7}Pb_{0,3}Sr_2Ca_2Cu_3O_{10-\delta}$ sample on the basis of precursors fabricated by the superfast melt quenching.

In summary, in high-temperature superconducting Bi-Pb-Sr-Ca-Cu-O samples, synthesized by the superfast melt quenching technology using solar energy, it was observed the “chaos” region, which could point to the fact of the existence of other, more high-temperature magnetic or superconducting phases, with $T_c \approx 119$ K in the matrix of the normal Bi(2223) phase with $T_c = 107, 5$ K.

REFERENCES

[1] M. Murakami, M. Morita, K. Doi et al. *Jap. J. Appl. Phys.*, 1989, V.28, №7, p.1189.

[2] D. D. Gulamova, D.E. Uskenbaev, G. Fantozzi, J.G. Chigvinadze, O.V. Magradze. *Zh. Tekhn. Fiz.*, 2009, v. 79, №6, 98-102.

[3] O. Eibl, *Physica* C, 1990, 168, 249-256.

[4] J. S. Luo, D. Michel and J.-P. Chevalier, *Revue Phys. Appl.* 1990, 25, 3-7.

[5] R J Lin, S W Lu and P T Wu, *Supercond. Sci. Technol.* 1989, 1, 336-339.

[6] Xiaohui Gao, Xiaolin Wu, Hui Yan, Zhoulan Yin and Caidong Lin, Yingsheng Fu, Wuxi Xie, *Modern Physics Letters* B, 1990, 4, №2,137-144.

[7] S.M.Ashimov, J.G.Chigvinadze, *Phys. Lett.* 2003, A313, 238.

[8] J.G.Chigvinadze, A.A.Iashvili, T.V.Machaidze, *Phys. Lett.* 2002, A300, 524.

In: New Developments in Material Science ISBN 978-1-61668-852-3
Editors: E. Chikoidze and T. Tchelidze

Chapter 3

SPECTROSCOPIC FEATURES OF LIF(OH) DUE TO IRRADIATION CONDITIONS

Z. Akhvlediani*, I. Akhvlediani, G. Dekanozishvili, T. Kalabegishvili and V. Kvatchadze
E.Andronikashvili Institute of Physics, Tbilisi 0177, Georgia

ABSTRACT

In the presented paper optical and paramagnetic spectra, besides that thermally stimulated luminescence (TSL) were investigated for LiF(OH) crystals irradiated at low (20 K and 100K) and at usual nuclear-pile temperatures. Some of the samples were irradiated in free or compressed states. It was established the dose dependence of the behavior of hydrogen defects formed after decay of hydroxyl ions under the action of ionizing radiation.

1. INTRODUCTION

The large variety of dosimeters based on LiF indicates that this material possesses unique properties as dosimetric substance. Moreover, LiF has very wide applications as tuning laser medium.

---

* E-mail: zairaak@yahoo.com

Lithium fluoride is a very convenient object for investigation of radiation defects formed in crystalline matrix under the action of ionizing radiation. That's why this material has been attracting the attention of the scientists for a long time.

Hydroxyl OH^- ions are one of the most extended impurities in ionic crystals, especially in LiF. In some cases even small amount of OH^- takes decisive role in radiation formation of defects and in their following evolution and stabilization. These impurity ions are often called the fundamental defects of crystal lattice.

At irradiation of hydroxyl-contained LiF crystals by ionizing radiation (neutrons, X- and gamma- rays) the different products of radiolysis are formed. During this process not only the host crystal but also the impurity presented in it are changed. Hydrogen products of radiolysis of OH^- impurity can be studied by the methods EPR and IR spectroscopy. Due to small masses, the hydrogen ions cause the appearance of high-frequency local oscillations in transparency region of crystal. In some cases the local oscillators are rather sensitive "probes" and by using them we can obtain information not only on the dynamics of defect lattice, but also on the distribution of impurities in crystal, on formation of complexes, etc. The nature of thermoluminescence (TL) in insulating crystals is still slightly blurred. Although it is known that irradiated samples of crystalline alkali halides, for instance, release luminous energy when heated above the irradiation temperature, the mechanisms of the phenomenon are still debated.

It should be noted, that the nature of crystal defects has been investigated over the past 60 years and in principle, TL measurements provide a very sensitive indication of complex defects in insulators. But the study of defect structures in solids using only this method is almost impossible. In fact, this technique has only limited possibilities in arriving at a true characterization of the solid under study. Therefore this type of measurement has been supplemented by many other experimental techniques.

The purpose of this work is a detailed study of total TL curves, optical and paramagnetic properties of irradiated LiF crystals containing OH^- impurity.

2. Experimental Procedures

Single-crystalline samples of lithium fluoride (20x10x5 mm^3) were cleaved out from blocks. The content of metallic impurities in these samples was determined by chemical analysis, hydroxyl ions - by IR spectroscopy. The content of Fe, Cr, Ni and Pb was less than 10^{-4}%, Mn and Cu even less by an order of magnitude.

As for the amount of magnesium it was less than 3×10^{-4}weight %. The hydroxyl impurity in studied crystals was only in "free "form [1] and composed about 180 ppm.

While some of the samples were irradiated by the 1.2 MeV gamma-rays of a ^{60}Co source, others were irradiated in nuclear reactor at low (20 and 100 K) and at usual reactor temperatures.

It should be noted that the mentioned procedure of irradiation was carried out about 25 years ago in Georgian nuclear reactor (Tbilisi). After IR spectroscopic studies the crystals were kept at room temperature. IR absorption spectra registered at present showed their identity with earlier ones, which indicates the stability of hydrogen defects formed in crystal.

TSL curves were registered in the regime of photon counting at temperature range 300 – 775 K with a constant rate of 1 K/s for heating the samples [2]. Luminescence was recorded integrally in the whole diapason of wavelengths by a photomultiplier. Optical absorption (electronic and vibration) spectra were studied at room temperature using spectrophotometers SF-26 and Specord-75 IR, respectively. EPR spectra were registered by radio-spectrometer RE 1306.

3. Results and Discussion

The investigation of the optical absorption bands of LiF continues to be of a great importance in the study of the TSL mechanisms in this material. Figure 1 shows the change of optical absorption spectra of LiF with the increase of irradiation temperature from 20 K up to reactor temperature. As one can see from the Figure at lowest irradiation temperature only the intensive F- and weak F_2-bands are observed. With the increase of irradiation temperature for the same irradiation dose the pattern becomes very complicated at the expense of formation of F-aggregate color centers.

It is well known that spectroscopic methods (measurement of absorption and /or emission spectra) for studying alkali halide crystals have no analogs. However, at high fluences of irradiation the mentioned spectra do not allow to make direct conclusion about behavior of formed defects, as the effect is very strong and pattern is highly complicated.

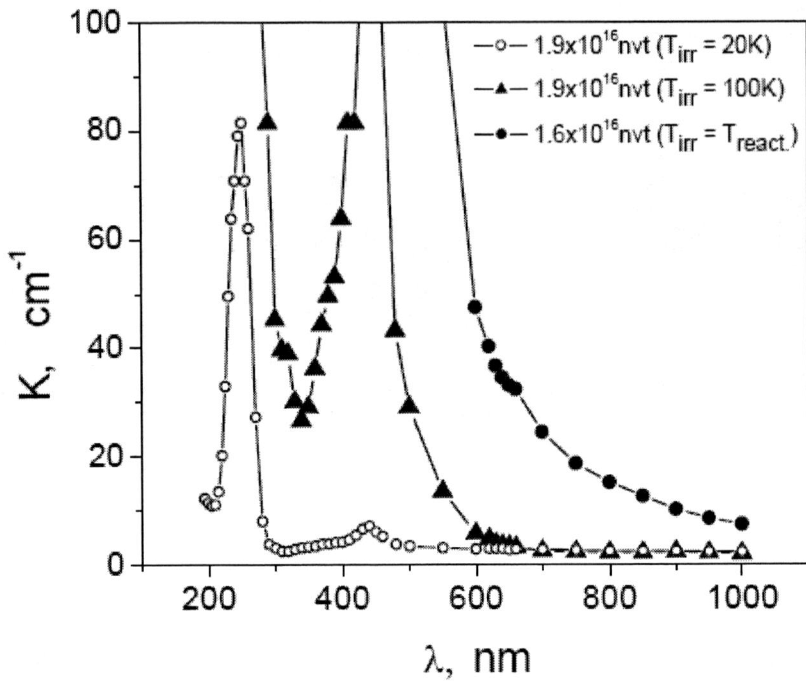

Figure 1. Evolution of optical absorption spectra of LiF with the increase of irradiation temperature.

We tried to "read" the radiation microstructure of the crystal under the conditions of thermal activation of the sample by TSL measurements. In a number of cases we had the unique samples, as nobody has studied the samples irradiated in reactor channel at 20 K. Besides that, we had the samples irradiated under loading (uniaxial compression in [100] direction at $9x10^6$ Pa) and in the free state (without loading).The study of such samples was also a great interest.

Figure 2 shows TSL curves for LiF irradiated in the reactor channel under different conditions (evolution of the processes taking place in the crystal is clearly seen in the Figure). In Figure 2, curve 1 corresponds to the lowest irradiation temperature 20 K. According to the curve there is almost no structure in the spectrum. It means that we have minimum lattice distortion in the sample. The increase of irradiation temperature (curve 2 – T=100K) causes the significant increase of TSL intensity even at lower dose. At the same time the structure of glow curves appears giving evidence of the variety of traps.

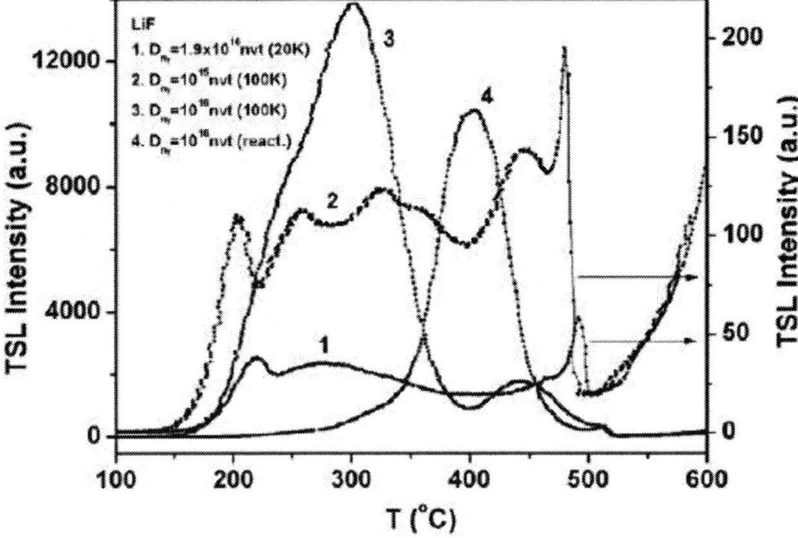

Figure 2. Evolution of glow curves of LiF under the action of n,γ-radiation.

The increase of dose when keeping the irradiation temperature constant (T=100K, curve 3) significantly increases the concentration of the formed defects. Taking into account that the process of formation of complex centers proceed intensively, it becomes clear that in this case we have practically one complex peak (if the second, much smaller peak is neglected), which is not characterized by the apparent structure. The simplicity of TSL spectrum becomes especially impressive.

Finally, curve 4 shows that at the increase of the irradiation temperature up to usual nuclear–pile temperature and at constant fluence (10^{16} n/cm^2) the previous peak shifts to the region of higher–temperatures. This is rather natural because in this case the "acting" defects are formed at higher irradiation temperatures, they are "survived" under these conditions and are thermally more stable.

Let us consider especially the results shown on the curve 2 in Figure 2. The first four peaks revealed by us up to 400°C are known from literature [3]. It should be only noted that the 1st and 2nd peaks on the curve are the 5th and 7th peaks well known in TLD dosimeters [4]. The third peak on this curve (332°C) and the main peaks on curves 3 and 4 (at 308 and 410°C, respectively) according to Baldacchini et al. (2008) should be assigned tentatively to F-like aggregate centres in the crystals [3].

On the curve 2 the most interesting are high-temperature peaks right to 400°C. If we take of into account the well-known data on annealing of F-centers [3], one can relate the peak at 450°C to the ionization of these centres.

Finally, the peak about 500°C: this high-temperature peak we relate to OH^- ions according to our publication [5] concerning to inter-transformation process of hydrogen centres and OH^- ions. At the temperature above 450°C hydroxyl begins to recover, i.e. these ions return to anionic sites of crystal which should be accompanyed by the emission of TSL quantum.

Investigation of the optical properties of irradiated LiF(OH) crystals in the near IR spectral region was first carried out in [6,7]. The observed absorption bands in the interval of wave numbers 1900-2200 cm^{-1} were attributed to the local vibrations of interstitial hydrogen ions H_i^- (U_1-centers). It was shown that these centers are stable up to 250°C. This was in contradiction with existing opinion that these centers can be stable only at low temperatures. Further on, Gellerman et al. in [8] gave the value of maximum frequency of local oscillations of U_1-centers in LiF being in good agreement with the results of [6, 7]. As it appeared lately, the shape and the spectral position of the IR absorption bands of U_1-centers in LiF strongly depend on conditions of irradiation.

For samples irradiated by gamma-rays under loading (uniaxial compression in [100] direction at $9x10^6$ Pa) and in a free state (without loading) we determined the dependence of concentration of U_1-centers on irradiation dose up to 12 Mrad (Figure 3).This dependence gives evidence of linear correlation between the number of interstitial hydrogen ions and irradiation dose (k $\Delta\nu$ – the IR optical band area, is proportional to the concentration of corresponding centers). This fact can be of a great importance as the most convenient method appearing for estimation of the values of ionizing radiation. And even more, this method does not erase the information stored in the sample. Just this property we hoped to find in TSL measurements, i.e. to reveal the behavior of U_1-centers in some TSL peaks, as these centers are the electron traps.

It is necessary to note that in the mentioned interval of gamma-irradiation doses the amount of other hydrogen products of OH^--decay (atomic hydrogen and substitution hydrogen ions, localized in anion sites of the crystal) has maximum at the dose about 5 Mrad (Figure 4a and Figure 4b, respectively). K_{max} – absorption coefficient in maximum of IR optical band for H_s^- ions. At the constant halfwidth it is proportional to the concentration of absorbing centers.

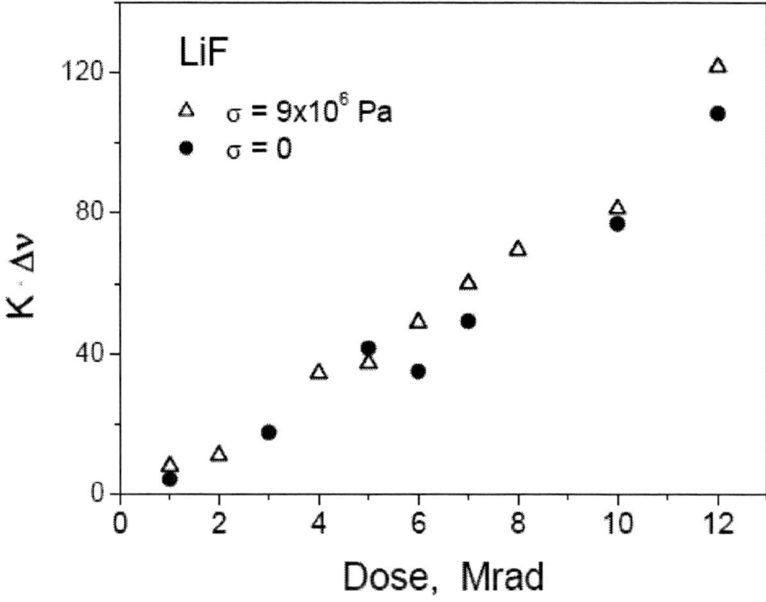

Figure 3. Linear correlation between the amount of interstitial hydrogen ions H_i^- and γ-irradiation doses (irradiation in stressed and free states).

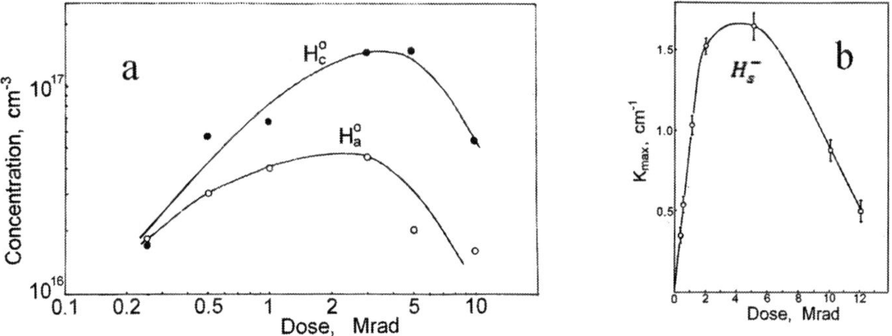

Figure 4. Variation of concentration of hydrogen atoms (a) and substitution ions H_s^- (b) with irradiation dose.

CONCLUSIONS

1. In TSL spectrum of irradiated LiF(OH) the glow peak was registered near 500°C (775K) that was not observed earlier. This peak is tentatively related to the process of restoration of OH^- ions during thermal annealing.
2. The linear correlation is determined between the number of interstitial hydrogen ions and the dose of gamma-irradiation for LiF(OH) in the interval from 1 to 12 Mrad. This correlation can be used for detection of gamma-radiation.
3. In the mentioned interval of doses the accumulation of atoms and substitution ions of hydrogen passes over the maximum.

REFERENCES

[1] Stoebe, T.G. *J.Phys.Chem.Solids*. 1967, vol. 28, 1375-1382.

[2] Kvachadze, V.; Dekanozishvili, G.; Vylet, V.; Galustashvili, M.; Akhvlediani, Z.; Keratishvili,N.; Zardiashvili ,D. *Radiat.Eff.Defects Solids*. 2007, vol.162 (1), 17–24.

[3] Baldacchini,G.; Chiachiaretta, P.; Gupta,V.;Kalinov, V.; Voitovich, A.D. *Fizika tverdogo tela*. 2008, vol. 50 (9), 1679-1686.

[4] Nepomniaschikh ,A.I.;Radjabov,E.A.;Egranov,A.V.; in "*Color centers and luminescence*",Novosibirsk, 1984, 51-72.

[5] Akhvlediani,Z.; Kvatchadze, V.; *phys.stat.sol.* (c). 2007, vol.4 (3), 1163-1166.

[6] Akhvlediani,Z.; Politov, N. *Izvestia* AN SSSR. 1971, vol. 35 (7), 1414-1418.

[7] Akhvlediani, Z.; Berg, K.-J.; Berg, G. *Crystal Lattice Defects*.1980, vol.8, 167-175.

[8] Gellerman, W.; Luty,F.; Koch, K-P.; Litfin, G. *phys.stat.sol.*(a) .1980, vol.57, 411-418.

In: New Developments in Material Science ISBN 978-1-61668-852-3
Editors: E. Chikoidze and T. Tchelidze

Chapter 4

PREPARATION OF $SmSb_2$ FILMS AND INFLUENCE OF VALENCE OF AN ION SM ON THE OPTICAL PROPERTIES

A. V. Gigineishvili, Z. U. Jabua* and M. S. Taktakishvili
Georgian Technical Univercity
77 Kostava St., Tbilisi 0175, Georgia

ABSTRACT

The effect of samarium ion valence on the optical properties of samarium diantimonide thin films has been studied. The optical constant spectra were obtained by processing of the reflectance spectra of $SmSb_2$ films at different values of samarium ion valence (+2.7; +2.4; +2.2) by the Kramers-Croning method. At all the values of samarium ion valence, there remained a gap in the energy spectrum and a clear plasma edge in the spectral dependence of reflection coefficient. The decrease in valence causes the decrease in plasma frequency and in the effective number of charge carriers contribution to plasma oscillation and also reduces the role of f-electrons in the formation of pf and fd hybrid states. The energy zone extremum is very sensitive to changes in the valent state of samarium ion as well.

Keywords: Film, Evaporation, Valence, Energy gap.

---

* Correspondng author: e-mail: Z.Jabua@hotmail.com

1. Introduction

Pnictides of rare-earth elements (REE) posses interesting, in many respects contradictory properties [1-3]. The problem of valence of rare-earth ion in pnictides and its effect on the physical properties is quit topical, as the usually supposed trivalent state does not always occur [4-7].

We studied the effect of valent state of samarium ion on the spectral regularities of optical parameters of samarium diantimonide films. The spectral dependences of reflection (R) end absorption (α) coefficients, of real (ε_1) and imaginary (ε_2) parts of dielectric constant and of the function of loss (Im ε^{-1}) were determined in the photon energy band of 0.05-5.5 eV. The basic properties and the energy state of revealed structures in all spectra were studied. Their transformation with changing the valence of samarium from +2.7 to +2.4 and then to +2.2 were analyzed. Probable corresponding changes in the energy dependence of the density of states, the electron mechanisms responsible for the formation of a certain set of spectral characteristic, were considered.

2. Experimental Results

The monophase textured crystalline films of $SmSb_2$ of 0.5-1.1 μm thickness were made by thermal evaporation from two independent sources of Sm and Sb [8]. The films had rhombic lattice structure (structure type $LaSb_2$) and the following lattice parameters: a=6.19 Å, b=6.03 Å and c=17.67 Å. The electrophysical parameters of films we measured at 300 K [9] are characteristic of many compounds with intermediate valence: resistivity - 10^{-5} Ohm·cm, Hall constant ~ $10^{-3}-10^{-4}$ cm^3/C, Hall mobility ~10^2 /V·s, Seebeck coefficient - ~1μV/k.

By selecting the production process parameters, golden, black and blue films were obtained. As the X-ray analyses of L_{III} spectra showed, in the golden film, the Sm ion valence was equal to ν=+2.7; in the black film to ν=+2.4; in the blue film to ν=+2.2.

The reflectance and absorption spectra were measured in wide optical range of 0.05-5.5 eV at room temperature (substrate material: leucosapphire and silicon). The optical constant spectra were obtained by processing the reflectance spectra of $SmSb_2$ films at different values of samarium ion valence by the Kramers-Kroning method.

3. Result and Discussion

The spectral dependence shown in figures 1-4 and the data presented in the table below allowed us to follow the transformation of specific features in the optical spectra when the value of samarium ion valence changed from ν=+2.7 to ν=+2.4 and then to ν=+2.2.

Reflection. The long –wave part of spectrum did not change, and at all the values of ν, high reflectivity (about 70%) with the boundary character of the relationship between reflection coefficient R and frequency ω was retained.

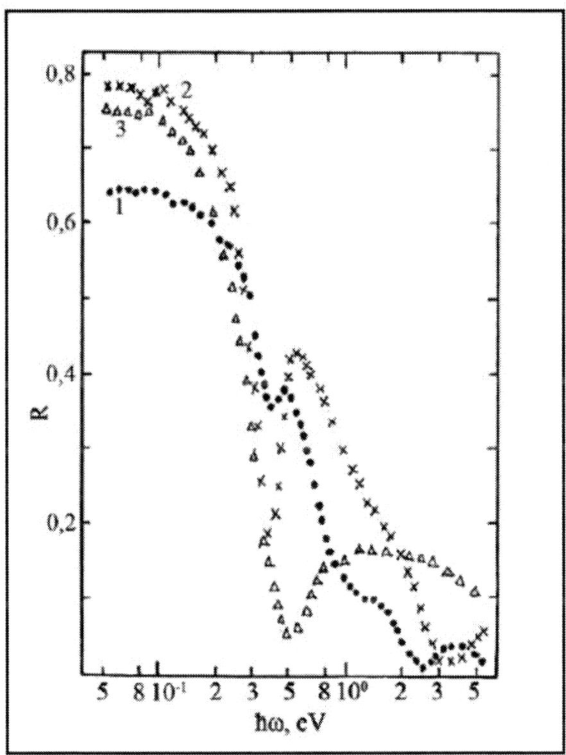

Figure 1. Spectral dependences of reflection of $SmSb_2$ with different values of Sm ion valence (1.- +2.7; 2.- +2.4; 3.- +2.2).

The rest spectral regions underwent significant transformation. As the valence decreased, the depth of minimum E'o increased; the reflection coefficient value dropped from 0.35 to 0.2 ev, an energy shift of 0.41-0.49 eV was also observed. The amplitude and the energy of corresponding signal maximum E_3 grew from 0.46 to 0.62 eV, and at v=+2.2, this maximum became smeared, merging with the reflection band which was extended over the range of 0.5-5.5 eV. In the short-wave spectral region, the changes were the most transient. Maximum E_5 and minimum E_0 underwent the „red shift" by 1.35 eV and 0.7 eV, respectively, when the Sm ion valence decreased to +2.4. Maximum E_5 was replaced by minimum E"o, but all these structures disappeared in the film where the samarium ion valence was equal to v=+2.2. Evidently the formation of a very strong mechanism of absorption, the contribution of which would become dominant over a wide range of photon energy, took place.

Absorption. At any valent state of samarium ion, in the $SmSb_2$ films, there exists a mechanism of absorption with characteristic energy (at a maximum) of 100-200 meV, while the minimum energy of transition is likely equal to 50 meV, which is out of the range under study. Maximums E_2,E_3 and E_4 in the absorption coefficients spectra point to the existence of a few peaks of density of states of different origin at $v > + 2.2$. The small half-width of these structures in the film with v=+2.7 should be considered as a consequence of high-degree of localization of energy state. Thus, it is evident that f-states and/or fd and fp hybrid states play an important role in the optical processes. Widening and even merging of the these absorption maximums as the samarium ion valence decreases point to a fast decrease of the degree of localization, to the corresponding energy widening of the maximum of density of states and, hence to reduction of the role f-electrons at formation of pf and fd hybrid states.

It can also be supposed that, at $v = + 2.2$, there is a direct gap equal to 0.6 eV between a wide conduction band and a more narrow, occupied band formed, apparently by the pf hybrid states. The value of absorption determined by transitions of electrons between these band in the visible spectrum is so high that the contribution of interband transitions from the standard p(sb)-valence band does not manifest itself in an explicit form. The energy of these transition can be estimated by the location of minimum E_0 in the spectra of α- and ε_2 –films with v=+2.7 and v=2.4, respectively. The instability of energy location of this minimum points to significant sensitivity of the valence of rare-earth ion. In the framework of the model under consideration, the band at ~2 eV could be either of exciton origin or determined by indirect transitions of electrons between the fp band and the conduction band.

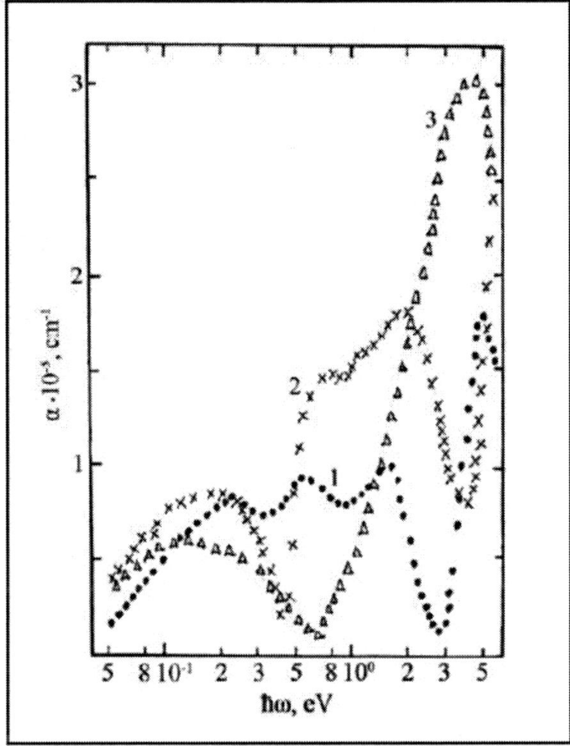

Figure 2. Spectral dependences of absorption of $SmSb_2$ films with different values of Sm ion valence (1.- +2.7; 2.- +2.4; 3.- +2.2).

Collective processes. High concentration of charge carriers in the compounds of intermediate valence determines an important role of collective oscillation processes in the formation of spectral dependences of optical parameters. By the contribution of plasmons, the long-wave edge of the reflection coefficient is usually determined, which, in its turn, used to be associated with metal-likeness of the compounds with variable valence [10]. However, in the works [11, 12], it was demonstrated that plasma frequency is determined by the effective number of electrons which all rare-earth ion contribute to. Thus, the variation of "plasma" parameters of optical spectra in a wide range becomes possible at a semiconductor approach to variable valence too (permanent existence of a gap in the energy dependence of the density of states is supposed) This fact is also justified by the results of our investigation.

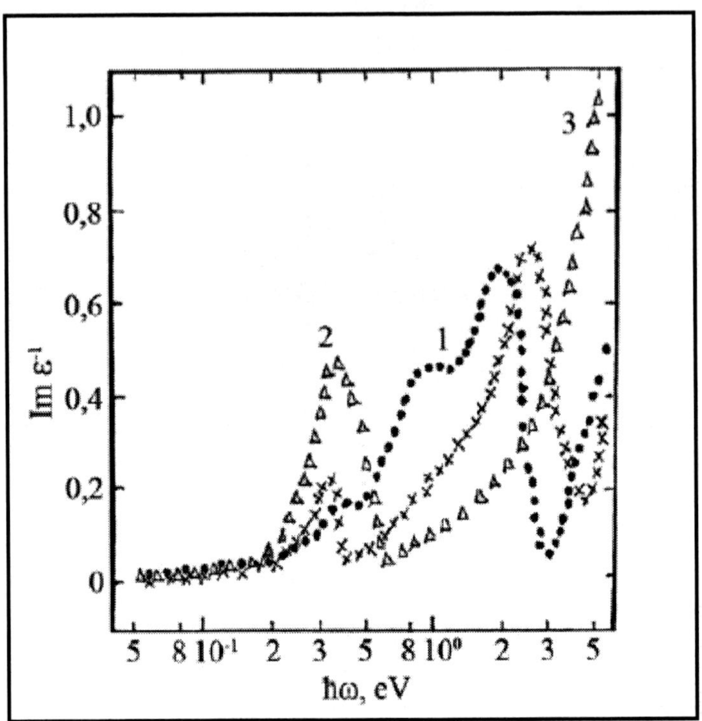

Figure 3. Spectral dependences of the function of loss of $SmSb_2$ films with different values of Sm ion valence (1.- +2.7; 2.- +2.4; 3.- +2.2).

At all the values of samarium ion valence, most likely there is a gap retained in the energy spectrum and, at the some time, the clear plasma edge in the spectral dependence of reflection coefficient. Judging by the energy location of E' (zeros of $\varepsilon_1(\omega)$), when the samarium valence decrease, only the plasma frequency and hence, the effective number charge curriers contributing to plasma oscillation decrease.

It should be noted that a good agreement between the E" values (0.34 eV and 0.36 eV) in the spectra of the real part of dielectric constant and of the function of loss (energy maximum $\mathrm{Im}\varepsilon^{-1}$) is achieved only if the relatively low value of valence causes weaker optical activity of interbend electron transition in the plasma frequency range.

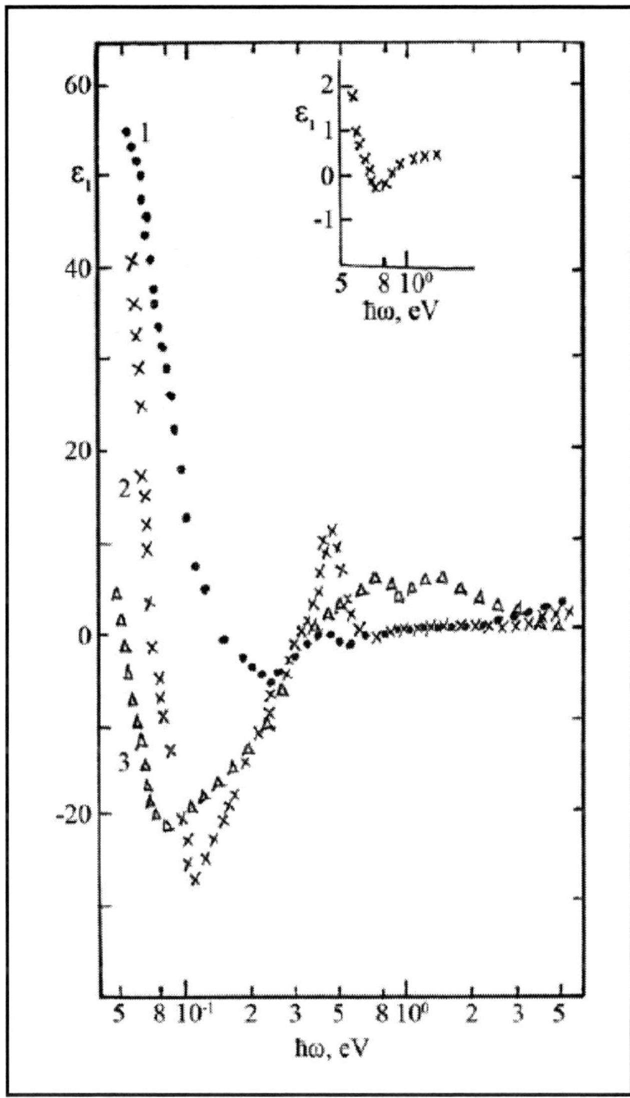

Figure 4. Spectral dependences of the real part of dielectric constant of $SmSb_2$ films with different values of Sm ion valence (1.- +2.7; 2.- +2.4; 3.- +2.2).

Table 1. Energy of the basic structures in spectra of a $SmSb_2$ films with various valent condition of an ion Sm (E' and E'' are the energy values corresponding to the intersection of the energy axis $\varepsilon_1(\omega)=0$ by the frequency dependence of the real part of dielectric constant with a negative and a positive inclination, respectively; E'_0 and E_0 are the minimum energy values of the parameters; E_1- E_5 are the maximum energy values of the parameters. Unit of energy- eV, unit of α - 10^5, cm^{-1})

Ion valence Sm	parameter	E_1	E'	E_2	E'_2	E'_0	E_3	E''	E_4	E_0	E_5	E''_0
+2.7	R					0.41	0.46		1.55	2.7	4.2	
	α			0.25			0.55		1.55	2.93	4.9	
	ε_2	0.08		0.14			0.44			2.93	4.37	
	ε_1		0.14			0.41		0.78				
	$Im\varepsilon^{-1}$					0.41		1.0	1.8			
+2.4	R				0.44		0.62			2.0	2.85	4.1
	α				0.38		0.78		1.67		3.0	3.95
	ε_2				0.44		0.59			2.0	2.75	3.95
	ε_1							0.34				
	$Im\varepsilon^{-1}$							0.34				
+2.2	R					0.49			1.4			
	α			0.14		0.60					4.3	
	ε_2					0.60			1.77			
	ε_1		0.55	0.11				0.34				
	$Im\varepsilon^{-1}$							0.36				

CONCLUSION

The transformation of specific features in the optical spectra of samarium diantimonide thin films showed that, when the Sm ion valence decreased, there was formed a very strong mechanisms of absorption, the contribution of which became dominant in a wide range of photon energy. It was supposed that, in the films with low valence of samarium ion (+2.2), there was a direct gap between the conduction band and the occupied gap. The analysis of experimental results demonstrated that at all the values of samarium ion valence, the gap was retained in the energy spectrum and, at the some time, the clear plasma edge appears in the spectral dependence of reflection coefficient.

REFERENCES

[1] M. Gasgnier. *Phys. Stat. Sol* (a). 114,11 (1989).

[2] G. Vaitheeswarn, L. Petit, A. Svane, V. Kanchana, M. *J. Phys. Condens. Matter*. 16,4429(2004) .

[3] M.N. Abdusalyamova. J. of D.I. Mendeleev chemical society. 26,673(1981) (in Russian).

[4] M.I. Stamateli. Abstracts of the international conference *„Spine Electronic: Novel Physical Phenomena and materials"*. Tbilisi, 2007, P. 56.

[5] R.D. Parcs. Mixed valence phenomena: an Overview. *Hyperfine Interaction*. 25, 565 (1985).

[6] L.N. Glurjidze, T.O. Dadiani, Z.U. Jabua, T.L. Plavinski, A.V. Gigineishvili, E.V. Dokadze, V.V. Sanadze, L.D. Finkelshtein, N.N. Efremova. Proccedings of the fourth Soviet-West Germany workshop. Sukumi. Tbilisi, 1988, P.167.

[7] E.P. Fetisov, D.I. Chomski. *J. Exp. Theor. Phys*. 92,105 (1987) (in Russian).

[8] Z.U. Jabua, E.V. Dokadze, L.N. Glurjidze. *Bulletin of the Academy of sciences of the Georgian SSR*. 138,101 (1990).

[9] A.A. Vinogradov, V.V. Kaminski, I.A. Smirnov. *Sov. Solid state Phys.* 27 ,1121(1985) (in Russian).

[10] B. Batlogg. *Phys. Rev.* B. 23 ,1827 (1981).

[11] K.A. Kikoin. *J. Exp. Theor. Phys*. 85,1000 (1983) (in Russian).

[12] G.P. Nijnikova, I.B. Nechludov, K.K. Sidorin, O.V. Farberovich. *Sov. Solid. State Phys*. 29, 803(1987) (in Russian).

Changes in samarium ion valence from +2.7 to +2.4 and then to + 2.2 was studied. It 4.

In: New Developments in Material Science ISBN 978-1-61668-852-3
Editors: E. Chikoidze and T. Tchelidze

Chapter 5

SPECTRA OF THE MAGNETIC VIDEO-PULSE INFLUENCE ON THE NUCLEAR SPIN ECHO IN MULTIDOMAIN MAGNETS

G. I. Mamniashvili[1*], T. O. Gegechkori[1], T. N. Khoperia[1], T. I. Zedgenidze[1], F. K. Akopov[1], A. M. Akhalkatsi[2] and T. A. Gavasheli[2]

[1]E. Andronikashvili Institute of Physics, 6 Tamarashvili st. 0177. Tbilisi, Georgia

[2]I. Javakhishvili Tbilisi State University, 3 Chavchavadze av. 0128. Tbilisi, Georgia

ABSTRACT

We present the first systematic study of timing and frequency diagrams of magnetic video-pulse influence on the NMR two-pulse echo in a number of magnets (ferromagnets, ferrites, half metals, manganites). It is shown that the dependence of two-pulse echo intensity on the temporal location of a magnetic video-pulse in respect to radio-frequency pulses and spectral diagram of this influence are defined mainly by local hyperfine field anisotropy and domain walls mobility. These diagrams could be used for the identification of the nature of NMR spectra in

* E-mail: gmamniashvili@yahoo.com

multidomain magnetic materials and to improve the resolution capacity of the NMR method in magnets.

The possibilities of using different methods of nuclear spin-echo spectroscopy for studying properties of magnetically ordered substances were analyzed in a large number of works [1]. One of such widely employed methods is based on the introduction of additional pulses of a dc magnetic field into the system of exciting radio-frequency (RF) pulses; these additional pulses were called video-pulses (VPs), since they lack the filling frequency. Thus, in [2–4] the VPs were used to investigate the properties of domain walls (DW) in the europium iron garnet $Eu_3Fe_5O_{12}$, in ferrites with a spinel structure, thin magnetic films, and $Y_{2-x}Gd_xCo_{17}$ compounds with the substitution of Gd for Y ions. In these works, the different role of VPs during their symmetrical and asymmetrical position with respect to the second RF pulse in the two-pulse echo (TPE) procedure was revealed. These differences make it possible to find the coercive-force-related distribution of DW upon the symmetrical arrangement of VPs, and the anisotropy of the hyperfine field (HF) at the nuclei in the case of its asymmetrical arrangement. In particular, the detection [3] of the inhomogeneous influence of VPs on different sections of the frequency spectrum of NMR made it possible to reveal those crystallographic positions that prove to be preferable upon the substitution of Gd for Y.

Previously, such a procedure was used [2] for determining the magnetic-field strength which shifts a DW by a distance equal to its thickness. The scheme of the experiment is shown in figure 1a., borrowed from [4].

The authors of [4] investigated the influence of VPs on the intensity of TPE. They showed that in lithium ferrite the maximum suppression effect is achieved when the VP coincides with one of the RF pulses. This is caused by the fact that the action of a VP on the multidomain ferromagnetic material is in essence reduced to a reversible (in weak magnetic fields) displacement of DW. Since the factor of enhancement of the RF field η is non-uniform inside the DW, the RF pulses in the case of symmetric magnetic VP action excite nuclei located in DW positions with different η. These results in the reduction of the TPE amplitude because it is maximal when nuclear magnetization turning angles of two RF pulses are equal [4]. For this reason when magnetic pulse coincides with a RF pulse the echo signal should be reduced as it is seen for lithium ferrite, Figure 1a.

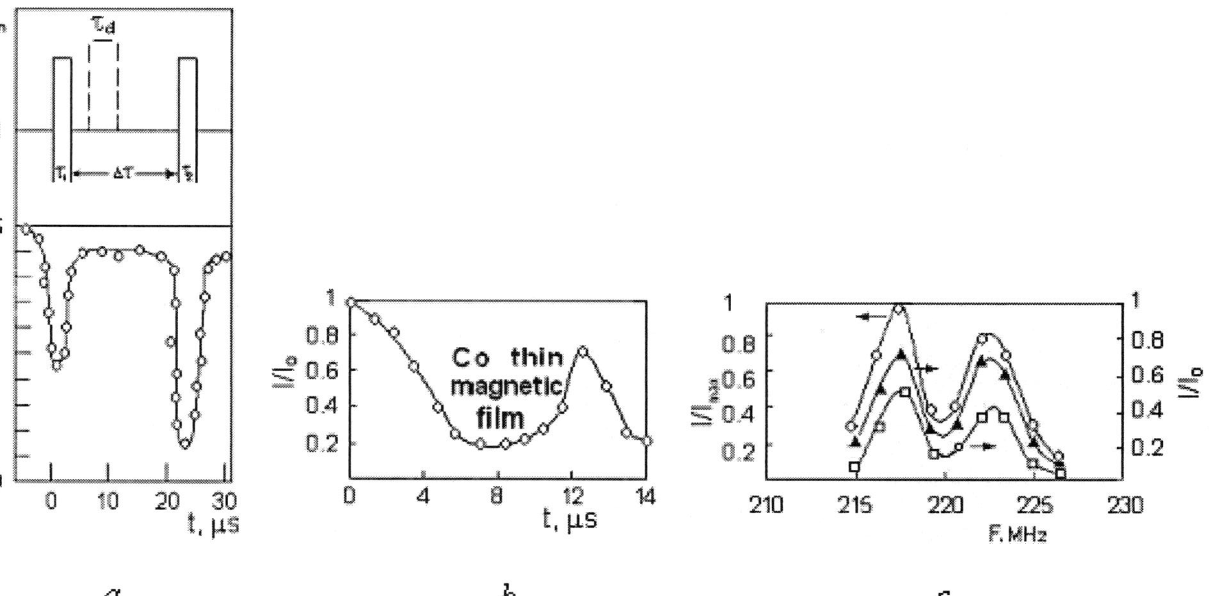

Figure 1. *a*) Timing diagrams of the relative intensity I/I_{max} dependence of the two-pulse echo (•) on the temporal location of a magnetic video-pulse of H_d=5 Oe, for ^{57}Fe NMR in lithium ferrite at: $\tau_1=\tau_2=0.8$ μs, $\Delta\tau=21$ μs, $\tau_d = 3$ μs, $f_{ЯМР}$=74.0 MHz, H_d=0. τ_1, τ_2, $\Delta\tau$, τ_d are rf pulse durations, time interval between them and magnetic pulse duration, correspondingly; *b*) Timing diagrams of the intensity dependence of the two-pulse echo on the temporal location of a magnetic video-pulse of H_d, in cobalt thin magnetic films at: $\tau_1 = \tau_2 = 1.5$ μs, $\Delta\tau = 9$ μs, $\tau_d = 3$ μs, H_d=10 Oe, f_{NMR}=216.5 MHz, I_o – echo amplitude at H_d=0; *c*) NMR spectrum of Co film (♦) and frequency spectra diagrams for magnetic video-pulse influence for symmetric (▲) and asymmetric (□) application at: $\tau_1 = 1.3$ μs, $\tau_2 = 1.5$ μs, $\Delta\tau = 9$ μs, $\tau_d = 3$ μs, H_d=10 Oe.

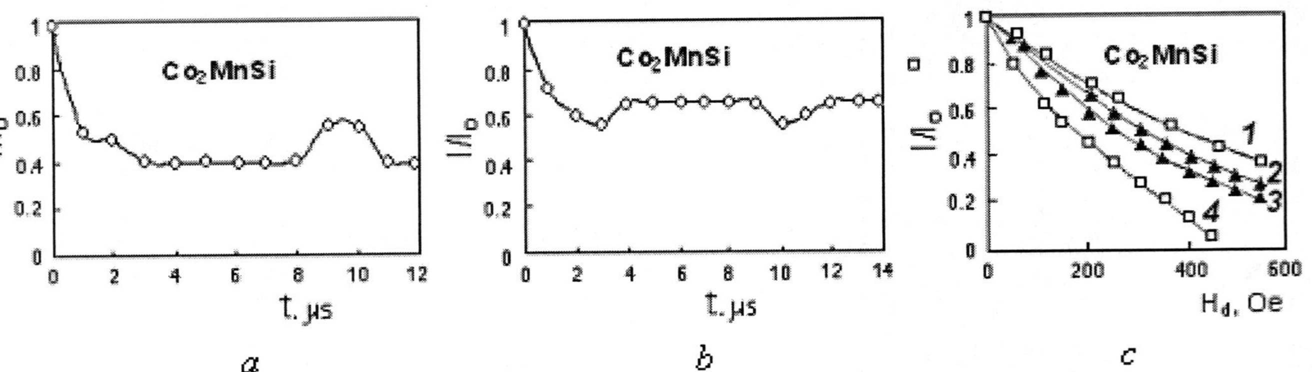

Figure 2. Timing diagrams of the intensity dependence of the two-pulse echo on the temporal location of the magnetic pulse with duration τ_d in Co_2MnSi (*a*) for ^{59}Co NMR at: $\tau_1 = 1.1$ μs, $\tau_2 = 1.4$ μs, $\Delta\tau = 10$ μs, $\tau_d = 2$ μs, f=145,5 MHz, H_d=550 Oe and (*b*) for ^{55}Mn NMR at: $\tau_1 = \tau_2 = 3$ μs, $\Delta\tau = 7$ μs, $\tau_d = 2$ μs, f=354 MHz, H_d=300 Oe. *c*) Two-pulse echo intensity dependences on magnetic video-pulse amplitude H_d in Co_2MnSi for ^{55}Mn for symmetric (▲) (2) and asymmetric (□) (1) influence at: $\tau_1 = 0.8$ μs, $\tau_2 = 0.9$ μs, $\Delta\tau = 8$ μs, $\tau_d = 1.6$ μs, f_{NMR} =353 MHz. and for ^{59}Co spin echo for symmetric (▲) (3) and asymmetric (□) (4) influence at. $\tau_1 = \tau_2 = 2$ μs, $\Delta\tau = 10$ μs, $\tau_d = 3$ μs, f_{NMR} =145 MHz.

The application of a VP asymmetrically in the interval between the RF pulses can influence the intensity of TPE only in the case of anisotropic HF interaction [2].

But here, of great importance may be characteristics of a magnetic material such as the mobility of DW and the HF anisotropy of HF interaction, which are different in magnetically soft cubic lithium ferrite and magnetically harder material such as uniaxial cobalt.

To study the above properties of the magnetic materials in more detail, we carried out experiments concerning the effect of a VP with an amplitude of the magnetic field up to $H_d = 500$ Oe and a duration equal to several microseconds on the signals of TPE in several magnetic substances, namely, polycrystalline Co, lithium ferrite $Li_{0.5}Fe_{2.35}Zn_{0.15}O_4$, and half metals Co_2MnSi, NiMnSb and manganite $La_{1-x}Ca_xMnO_3$ system. The half metals and manganites are regarded as promising materials for spintronics [5,6]. We present only results for cobalt thin magnetic film (TMF), Co_2MnSi half metal and $La_{1-x}Ca_xMnO_3$ manganite system. The procedure of the preparation of samples and magnetic susceptibility measurements is described in [4,7-10].

As it follows from Figures 1–4, the essential difference takes place for magnetic VP action on the TPE in different magnets. The dependence of its intensity on the time of application of a VP for cobalt TMF and ^{59}Co echo signal in Co_2MnSi (Figure 1b,c and 2a), correspondingly, differs considerably from closer-to-each-other dependences for lithium ferrite (Figure 1a) and ^{55}Mn echo signal dependence in Co_2MnSi (Figure 2b,c) and manganites (Figure 3a,b; 4a). The influence of VPs on the frequency spectra of echo signals was studied in [2,3] where it was noted that this effect also permits one to judge the quality of a magnetic material investigated. The spectra of TPE under the influence of VPs for our TMF samples are given in Figyure 1c, and for manganites in Figures 3, which correspond to the symmetrical and asymmetrical positions of VPs with respect to the second RF pulse. As can be seen from Figures 1c and 4a,b, the VPs noticeably affect the echo spectra. However, in order to determine to which extent these effects are suitable for controlling the quality of magnetic materials, additional studies are required.

Note the difference in the timing diagrams for the magnetic nuclei ^{59}Co and ^{55}Mn in Co_2MnSi (Figure 2a,b). In our opinion, this indicates the difference in the anisotropy of HF fields for these nuclei. This difference substantially affects the dependences of the intensity of the TPE on the time of application of VPs: the VPs which coincide in time with the RF pulses less suppress the signals of echo for ^{59}Co nuclei and comparatively more reduce TPE of ^{55}Mn nuclei.

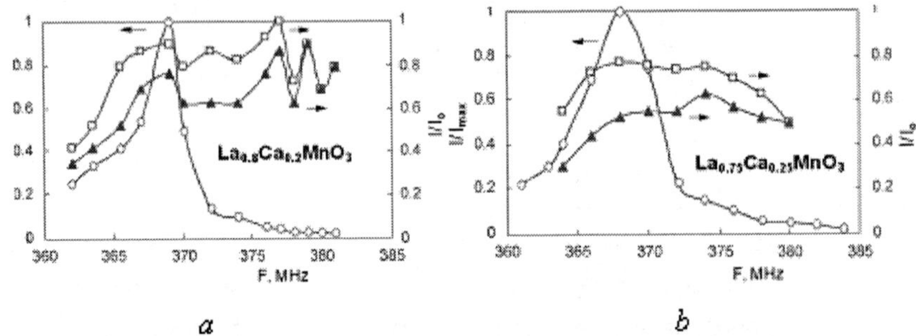

Figure 3. Frequency dependence of the effect of the symmetric (▲) and asymmetric (◻) magnetic video-pulses on two-pulse echoes (♦) at: $\tau_1 = \tau_2 = 2$ μs, $\Delta\tau = 10$ μs, $\tau_d = 3$ μs, H_d=50 Oe for ^{55}Mn NMR in (*a*) $La_{0.8}Ca_{0.2}MnO_3$ and (*b*) $La_{0.75}Ca_{0.25}MnO_3$.

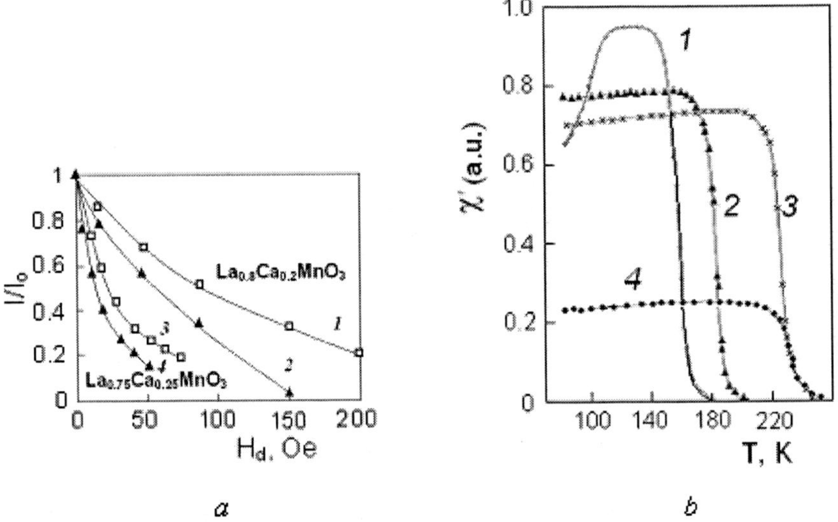

Figure 4. *a*)Two-pulse echo intensity dependences on magnetic video-pulse amplitude H_d for symmetric (▲) (Curve 2) and asymmetric (◻) (Curve 1) influence in $La_{0.8}Ca_{0.2}MnO_3$ and $La_{0.75}Ca_{0.25}MnO_3$ (curves 3,4) at: $\tau_1 = \tau_2 = 0.9$ μs, $\Delta\tau = 7$ μs, τ_d= 1.6 μs, f_{NMR} =368 MHz. *b*) Temperature dependence of the magnetic susceptibility for samples $La_{1-x}Ca_xMnO_3$: 1) x =0.2; 2) x =0.25; 3) x =0.1; 4) x =0.4.

So, the type of timing or frequency diagram is defined mainly by the HF field anisotropy of corresponding nuclei which is small for ^{57}Fe and ^{55}Mn as compared for ^{59}Co positions.

The reason for this it could be understood also from Figure 2c, where reduced echo intensity dependences on the magnetic VP amplitude is shown for ^{55}Mn and ^{59}Co positions in Co_2MnSi. It is seen that for these positions amplitude dependences for asymmetric action of magnetic VP differ much stronger then ones for symmetric influence.

So, the timing diagrams type is defined mainly by the anisotropic part of HF interaction while the degree of suppression of echo signals by magnetic VP strongly depends on the DW mobility.

Let us note also that timing diagrams in case of half metallic Co_2MnSi give a visual picture describing different HF anisotropies at the ^{59}Co and ^{55}Mn sites. In last years the $La_{1-x}Ca_xMnO_3$ system became a subject of numerous investigations due to the colossal magnetoresonance effect [6] revealed in it near the temperature of manganese spin magnetic ordering and also due to other observed interesting physical phenomena including the internal inhomogeneous spin states, phase separation, charge and/or orbital ordering [11,12]. It was revealed in this system that the transition from ferromagnetic state with semiconductor conductivity type to the ferromagnetic state with the metallic type conductivity at increasing the concentration x of doping is related with the redistribution of the volumes of two phases (two type of regions) differing by the degrees of electron hole mobilities at manganese sites. The NMR dynamics of nuclear spins points to the internal inhomogeneity of each of these phases. It is possible that the comparatively inhomogeneous frequency spectrum of magnetic VP influence in the sample with x=0.2 as compared with x=0.25 (in our experimental conditions signals at x=0.1 and x=0.4 were weak and was not studied.) and lower mobility of DW in this sample are related just with the above pointed phase state inhomogeneity characterized by a strong spread of DW mobility.

In summary, in this work it is carried out the first systematic study of timing and frequency diagrams of magnetic VP influence on TPE in a number of magnets with different anisotropy of HF fields and DW mobilities. It is shown that these timing and frequency diagrams are defined by local HF field anisotropy and DW mobilities and could be used for additional identification of the nature of NMR lines in multidomain magnetic materials and thereby to improve the resolution capacity of the NMR method for magnets.

ACKNOWLEDGMENTS

We are grateful to Prof. A.Shengelaya for kindly providing us with manganite samples. The work is supported by the Georgian National Science Foundation N GNSF/ST07/7-248 Grant.

REFERENCES

[1] M. I. Kurkin and E.A. Turov. NMR in magnetically ordered substances: Theory and applications. M.: *Nauka*. 1990.

[2] L.A. Rassvetalov and A.B. Levitskii. *Fiz. Tverd. Tela* 23 (11), 3353–3359 (1981).

[3] E. Machowska and S. Nadolski. *Solid State Commun.* 68 (2), 215–217 (1988).

[4] I.G. Kiliptari and V.I. Tsifrinovich. *Phys. Rev. B: Cond. Matt.* 57, 11554–11564 (1998).

[5] J.S. Moodera. *Phys. Today* 54 (5), 39–44 (2001).

[6] E. Dagotto, T. Hotta, A. Moreo. *Physics Reports* 344, 1 (2001).

[7] A.M.Akhalkatsi, T.O.Gegechkori, G.I.Mamniashvili, Z.G.Shermadini, N.A.Pogorely, O.M.Kuzmak. *Phys. Met.Metallogr.* 105, 351 (2008).

[8] T.N. Khoperia. Microelectronic Engineering, 2003, 69, 391; T.N. Khoperia. *Microelectronic Engineering*, 2003, 69, 384.

[9] G.M. Zhao, K. Cender, H. Keller, K.A. Müller. *Phys. Rev.* B 2000, 62, 5334-5336.

[10] F.K. Akopov, N.A. Arabadzhian, N.D. Kvataya, G.S. Oniashvili, I.V. Chkhartisvili. *Phys. Met. Metalogr.* 2004, 97, 3, 47-49.

[11] J. Dho, I. Kim, S. Lee, K.H. Kim, H.J. Lee, J.H. Jung, T.W. Noh. *Phys. Rev.* B59, N1, 492-496 (1999).

[12] G. Paparassiliou, M. Fardis, M. Belisi, T.G. Maris, G. Killias, M. Pissas, D. Niarchos, C. Dimitropoulos, J. Dolinsek. *Phys. Rev. Lett.* 84, N4, 761-764 (2000).

In: New Developments in Material Science ISBN 978-1-61668-852-3
Editors: E. Chikoidze and T. Tchelidze

Chapter 6

THE ELECTRONIC MECHANISM OF MELTING OF SEMICONDUCTOR MATERIALS

Z. V. Jibuti*, N. D. Dolidze and B. E.Tsekvava
Iv.Javakhishvili Tbilisi State University
13 I.Chavchavadze ave, Tbilisi, Georgia, 0179

ABSTRACT

It is shown experimentally and proved theoretically, that the reason of low temperature melting of a crystal at pulse laser annealing is the electronic mechanism of attenuation and isotropization of chemical bonds between its atoms. The formula is received for critical concentration of anti-bonding quasi-particles - electrons in conduction band and holes in valence band, above which a low temperature melting of the semiconductor on the electronic mechanism should be occurred. Possible mechanisms of non-thermal melting of semiconductors on an example of crystals with a covalent bond are discussed.

1. INTRODUCTION

Pulse photon annealing has been widely applied to electronics technologies for already several decades. The development of nanoelectronics

* E-mail: nugo42@mail.ru

made stricter requirements connected with the minimization of operating areas of semiconductor structures. In these conditions it is especially actual to search ways of carrying out of processes of modification of materials properties at relatively low temperatures, in order to prevent diffusion of impurity in undesirable area. The pulse photon annealing is considered by most authors as a short-time, but nevertheless a high-temperature heating of a material (up to temperatures of melting).

However, there are a number of experimental works testifying a significant role of athermic factors in these processes [1-8]. For example, the following experimental results demand an explanation:

- observed at photostimulated pulse laser annealing (PLA) of structural defects in Si [1,2]
- recrystallization process of amorphous GaAs by ionic implantation can be carried out at temperatures below the room temperature [3];
- recrystallization process of amorphized Si by ionic implantation is more effective for initial temperatures T_i=77K, than at T_i=300K [5,8];
- in InSb PLA it is equally effective, both at T_i =77K, and at T_i=300K [6];
- effective annealing of radiation defects created in GaAs by irradiation of accelerated electrons is absent [9,10].

The offered mechanisms of pulse photon annealing still remain disputable. Theoretical models of pulse photon (laser) annealing, which available in the literature, can be conditionally divided into two groups: the thermal mechanism-considering that the energy of a light pulse is transferred to lattice as during thermalisation of photo-excited carriers leading to a heating pre-surface layers and its melting with the following recrystallization [11,12], and models assuming that pulse laser annealing as not of a thermal nature and is connected to the effects caused by excitation in the pre-surface layers with high concentration of electrons and holes [4,13-15]. In these works the hypotheses for low-temperature PLA of semiconductors based on the concept of the change of a quantum state of valence electrons of a chemical bonds were proposed. The electron transition from valence band (bonding state) to conduction band (antibonding state) results in the birth of two antibonding particles - an electron in conduction band and a hole in valence band. Consequently the modification of force and the type of bonding (isotropization of bonds) take place. (For example, in semiconductors with a diamond structure (Ge, Si, etc.) and in III-V compounds (GaAs, etc.) the electron transition from bonding p - state in valence band (directed bonding) to antibonding symmetric state s - in conduction band). Within this model it is completely natural the existence of some critical concentration n_{cr} such that, if concentration of antibonding quasi-particles $n > n_{cr}$, the initial state of the semiconductor becomes unstable, depending on the conditions of the experiment - the power and the duration of PLA, low-temperature melting, accelerated diffusion, recrystallization of the amorphous layer, a structural phase transition into a metal state and etc may start.

A rigorous theory of electronic mechanisms of PLA is a complicated task of the nonequilibrium statistical physics and thermodynamics. Such an approach with some simplifying assumptions was realized in work [14, 15] for the covalent semiconductor - metal state structural transitions. For Si and GaAs $n_{cr} \approx 10^{21}$ cm^{-3} and $(7 \div 9) \cdot 10^{21}$cm^{-3} were obtained, respectively. According to the authors, such high levels of excitation can be achieved only at pico-femtosecond pulses of a laser.

A less rigorous but deserving attention approach was applied by the authors [4]. According to this work n_{cr} is such a value of the concentration of antibonding quasi-particles at which their local weakening actions are replaced by collective effect. Namely, it is supposed that n_{cr} is carrier concentration at which antibonding particles have time to be near each atom during one oscillation of lattice atoms. Consequently attenuation and isotropy of bonds take place simultaneously in the whole crystal, and a crystal - liquid phase transition occurs.

2. Theoretical Mode

Staying within the framework of the concept of a collective weakening action of the antibonding quasi-particles, an approach to derivate of the formula for estimates of n_{cr} other than presented in [4] was proposed earlier [16]. Opposed to bonding electrons in a covalent bond when a pair of electrons is collectivized between two neighboring atoms, the antibonding quasi-particle weakens bond not locally at any lattice point, but in entire region of its delocalization, i.e. in the region with the linear dimension of the order of its de-Broglie wavelength $\lambda_D = \dfrac{h}{p}$, (h – the Planck constant, p – the pulse of the particle). Assuming the volume of the sphere with the diameter λ_D is $V_D = \frac{\pi}{6}\lambda_D^3$. If the concentration of quasi-particles n is such, that for V_D there is one quasi-particle, then $nV_D \geq 1$. At such concentration in entire volume of the crystal, the antibonding quasi-particles weaken the bonds between all atoms. In this case the crystal can be considered as if it were a gigantic molecule with the weakened bonds between all its atoms.

However, "melting by the electronic mechanism" take place only if the level of weakening and isotroping of bonds reachs a certain critical value. Here we introduce our basic hypothesis: proceeding from the experimental fact, that melting is the result of weakening and isotroping of bonds between all atoms of the substance, we assume that the concentration of the antibonding quasi-particles is critical if the following realized

$$\mathrm{n}_{\mathrm{cr}}\, V_D(T_{mel}) = 1, \quad \mathrm{V}_\mathrm{D}(T_{mel}) = \frac{\pi}{6}\lambda_\mathrm{D}^3(\mathrm{T}_{\mathrm{mel}}) \tag{1}$$

It Means, that melting by the electronic mechanism occurs at such concentration of antibonding quasi-particles when in the spherical volume of the crystal of diameter equal to the de-Broglie wavelength at the melting point (T_{mel}), there is at least one quasi-particle. It should be emphasized, that λ_D (T_{mel}) and V_D (T_{mel}) are fixed, as formally designated length and volume is not connected with the real temperature of a crystal at "melting by the electronic mechanism". For a simple isotropic law of quasi-particle dispersion within the framework of classical nondegenerate statistics $p = \sqrt{3mkT}$ (m – Effective mass of conductivity of a quasi-particle, T – the absolute temperature, k - Boltzmann's constant). Then from equation (1), we obtain:

$$(n_{cr})_{c,v} = \frac{9\sqrt{3}}{\pi(2\pi)^{3/2}} N_{c,v} \approx 0.3N_{c,v}; \tag{2}$$

$$n_{cr} = 0.3N_{c,v} \approx 4.82x10^{15}(m_{c,v})^{3/2}\, T_{mel}^{3/2}\ (\mathrm{cm}^{-3}) \tag{3}$$

$N_{c,v}$ - effective density of states in conduction band (c) and in valence band (v) accordingly, m_c and m_v – the effective masses of conductivity of the electron and the hole in the units of a free electron mass, respectively. It should be noted that generally, the effective density of states in zones $N_{c,v}$ known in physics of semiconductors, differs from above mentioned value(in the formula (3) there the effective mass of conductivity, not the effective mass of density of states in the bands) is given. Coincidence takes place only for the scalar effective mass and of one of the case extreme in bands. Therefore while estimating n_{cr} it is necessary to take into account this circumstance.

At the simultaneous presence of both types of carriers, in the assumption of equality of their concentration (band - band transitions), for n_{cr} the following formula is received

$$n_{cr} = \frac{(n_{cr})_c (n_{cr})_v}{(n_{cr})_c + (n_{cr})_v} \tag{4}$$

Concerning formulas (1) - (4) it is necessary to make the following notes:

1. Above when defining the de-Broglie wavelength we used classical statistics. As it is known, the equilibrium concentration of electrons n_{oe} in conduction band and holes n_{op} in valence band, in the case of validity of the classical non-degenerative statistics, are expressed by the following formulas

$$n_{oe} = N_c e^{\frac{E_F - E_c}{kT}}, \quad n_{op} = N_v e^{\frac{E_v - E_F}{kT}} \tag{5}$$

Here, E_c and E_v – are the energies relevant to the bottom of the conduction band and the top of the valence band, respectively; E_F - the Fermi level. If the concentrations of quasi-particles are $n_e \leq N_c$, $n_p \leq N_v$, (which is fulfilled for critical concentrations (2)) according to (5) it means that the Fermi level is in the band-gap, and the nondegenerate statistics is approximately applicable. It is interesting to note that, for the product of critical concentrations we have

$$(n_{cr})_e \cdot (n_{cr})_v \approx 0.09 N_c N_v$$

Which can be formally presented as "the mass action law" for critical concentrations :

$$(n_{cr})_e \cdot (n_{cr})_v \approx N_c N_v \, e^{-\frac{E_g}{kT_{mel}}}$$

with a band-gap reduced to $Eg\sim(2\div3)\,kT_{mel}$

2. In formulas (1) - (5) it is assumed, that in the volume $V_D(T_{mel})$, there is one quasi-particle. However, for crystals of a diamond and zinc blende types with a tetrahedral base unit with four bonds, it is obviously reasonable to extend our hypothesis to presence of $i = 1\div4$ particles in $V_D(T_{mel})$. As a result, in formulas (2) and (5), the adding factor i will appear: $(n_{cr})_{c,v}\approx 0.3iN_{c,v}$.

The values of m_c, m_v and T_{mel} are: for Ge - 0.17, 0.2 and 1210K; for Si - 0.26, 0.37 and 1688K; for GaAs - 0.067, 0.34 and 1511K, accordingly. By using these parameters and formulas (2)-(4), the calculated values $(n_{cr})_{c,v}$ are (extreme values of n_{cr} are relevant to i=1 and i=4): for Ge - $(2.2\div9.6)10^{18}$ $sm^{-3}(n_{cr})_{c,v}$, for Si - $(0.8\div3.3)10^{19}$ sm^{-3} and for GaAs - $(1.4\div5.5)10^{18}$ sm^{-3}.

3. Experimental Mode

As a proof of validity of the stated model can serve experiments on PLA, which allows one to show an opportunity of realization of melting at a laser pulse energies below a threshold of thermal melting and dependence of efficiency PLA on the concentration generated by light antibonding quasi-particles. In the works [17-20] the technique is offered, where flowing in of the corrugated surface microrelief of a semiconductor material (GaAs, Si) at PLA is interpreted as the direct proof of melting of a surface.

By using the mentioned technique, we have carried out the following experiments:

On wafers of GaAs ($n_o=1\cdot10^{14}sm^{-3}$) the corrugated surface has been received. Samples were divided onto two groups. One of them remained without an irradiation (initial), and the second - was irradiated by accelerated electrons (E=3.0 MeV, $\Phi=5\cdot10^{16}sm^{-2}$). Samples of both groups were exposed to pulse laser influence (λ=0.69mcm, $\tau_{puls}=4\cdot10^{-4}$s, E=4 J/sm^2) with non-uniformity of energy distribution on an irradiated surface of ≤5 %. PLA was carried out, as in the air (T_i=300 K), as well as in the environment of liquid nitrogen (T_i=77 K).

The experiments showed, that:

- The flowing in of the corrugated surface at PLA was observed only in samples not subjected to an electron irradiation.
- In samples irradiated by electrons, melting was not observed.

- The effect of “flowing in the corrugated surface” is identical at both initial temperatures of crystal (T_i=300 K or T_i=77 K). Thus, the power threshold of “flowing in” (values of density of energy of the laser radiation, necessary for melting of a surface) does not depend on initial temperature of a laser irradiation.

The presence or absence of "melting" at PLA, observed in the experiment, can be connected with such factors as: thermal heating, electronic excitation, and in experiments in the environment of liquid nitrogen also with pressure pulse [6].

Within the model of thermal PLA estimations show, that in the conditions of the experiment the heating temperature of crystal is much less than the melting temperature.

Thus, we obtained:

- Regimes of the used PLA cannot provide achievement of the melting temperature of a crystal;
- In samples, irradiated by electrons, where the growth of temperature at PLA should be more (in view of decrease of factor of heat conductivity as a result of an irradiation by electrons [11, 12, 21]), melting is not observed;
- The power threshold of melting of initial crystals observed in the experiment is identical at both initial temperatures.

Estimations of influence of pressure pulse upon the change of value of the melting temperature of a crystal at PLA in the environment of liquid nitrogen (by the technique presented in the work [6]) have shows, that it is not significant ≈7K.

As seen from the above mentioned fact it is impossible to explain the observed effect of laser melting by pure thermal heating of crystal.

4. Discussion

Let's pass to discussion of experimental results within the model of electronic mechanism of melting. We shall estimate the concentration of the electron-hole pairs, generated by light.

Estimations [22] show that in the conditions of the experiment (energy of a laser pulse and its duration) in GaAs, in the rate equation for superfluous concentrations of electrons and holes, generated by light at a band-band transition ($\Delta n = \Delta p$), it is possible to neglect current members (transport of carriers), and also Auger-processes. As a result we have:

$$d\Delta n/dt = G - \Delta n/\tau - B(\Delta n)^2 \qquad (6)$$

where G - generation rate of electron-hole pairs, B - factor of radiation recombination at a band-band transitions ($B=2\cdot10^{-10}sm^3s^{-1}$ [22]). The estimation of a lifetime of superfluous carriers remained in (6) recombination mechanisms gives a value much less than the duration of a light pulse. It means, that on experiment stationary generation ($d\Delta n/dt=0$) is realized, and for Δn from (6) it follows

$$(\Delta n)^2 + \Delta n/B\tau - G/B = 0\,,\ \Delta n = \frac{1}{2B\tau}\left(\sqrt{1+4B\tau^2 G} - 1\right) \qquad (7)$$

For the estimation of concentration of the nonequilibrium carriers, generated by light, there such parameters as: factors of absorption (α) and reflections (R) on wavelength of laser radiation, by standard techniques; a lifetime of nonequilibrium carriers (τ), caused electrons and holes recombination through the impurity centers were measured, by the methods of investigation of kinetic of intrinsic photoconductivity and measurement of time of restoration of back resistance.

Using measured values R, α and τ, we receive:

For initial samples (R=0.3, $\alpha=3\cdot10^4sm^{-1}$, $\tau=8\cdot10^{-7}s$) $\Delta n_1 \approx 2\cdot10^{18}sm^{-3}$;

For irradiated by electrons samples (R=0.3, $\alpha=5\cdot10^4sm^{-1}$, $\tau=2\cdot10^{-10}s$) $\Delta n2 \approx 2.5\cdot10^{17}sm^{-3}$.

As seen from given above values $(n_{cr})_{c,v}$, in the initial samples stationary concentration generated by light quazi-particles n_{cr} is achieved and "cold melting" of a crystal (a flowing in of corrugated surface of a crystal) by the electronic mechanism should be observed, and in samples irradiated by electrons, where $n < n_{cr}$, the effect of “flowing in of corrugated surface” should be absent, as observed in the experiment.

From the above mentioned values $(n_{cr})_{c,v}$, can be seen for one antibonding quasi-particle there is 10^3 bonds are weakening. The questions are arisen: what is a mechanism of such collective weakening of bonds? Why do electrons of the conduction band and holes of the valence band weaken bonds between atoms of a crystal?

Because of the big mathematical difficulties, it is hardly possible to construct a strict quantitative theory of these phenomena, proper to the multielectronic problem. We shall try to answer these questions qualitatively, on an example of covalent crystals, basing on the two following models:

1) As known, covalent bonds "are localized" between atoms of a crystal and realized by paired electrons i.e. by electrons in one state with the opposite spins. We shall consider interaction of a band electron with any of these paired electrons. Digit 1 designates a band electron, and digits 2 and 3 designate bond electrons. Spin of the band electron should be parallel to spin of one (let us suppose 2), and antiparallel to another one from the paired electrons. Spin wave function of such system should be described by Young's scheme, in which it should be antisymmetric in relation to the rearrangement of a spin of bond electrons 2 and 3 (a singlet state) and symmetric concerning rearrangement of spin variables of band electrons 1 and 2. To satisfy Pauli principle symmetry of coordinate wave function of system should be defined by so-called transposed Young's scheme. This scheme describes coordinate wave function, which is antisymmetric to coordinates of band electron 1, and bond electron 2. But antisymmetric by spatial coordinates wave function corresponds to the situation, when they rarely get between positive nucleus because of the correlations of movement of 1 and 2 electrons. As a result, band electron 1 weakens bonds due to the exchange interaction with bonds electrons.

2) In the following model the answer to the risen questions, is based on the analogy to a problem of inelastic scatering of an electron on Para helium. Electrons of Para helium represent an analogue of paired electrons of bond, and a band electron - scatered electron, with some additional consideration of scatering process. Being based on a principle of identity of electrons, we assume, that electron of the conduction band (it is antibonding quasi-particle) replaces one of the electrons bonding orbital and settles with the "remained" electron with parallel spin, transforming a connecting orbital into a loosing one (break of bond), and a liberated electron does the same thing with the following bond and etc. Similarly, a moving hole of the valence band represents a broken bond, and an electron, having been occupied a place of the hole (movement of the hole) can transform a new orbital into a loosing one,

"having sat" on empty bond with the remained electron having a parallel spin. At presence of both types of quasi-particles both mechanisms are carried out simultaneously. We assume, that the loosing orbital is occupied by the electron during a very short interval of time ($\Delta t \approx 10^{-15}$s. – is of the order of the electron inter-nuclear distance flight time of - time needed states for exchange), and it is an intermediate virtual state of electron. According to the uncertainty principle for energy we have: $\Delta E \sim \hbar/\Delta t \approx 1eV$. Therefore, the energy is not conserved at transition of the electron to an loosing orbital. Its conservation takes place for the process as a whole, i.e. for a chain of transitions: band electron → an intermediate virtual state → band electron. (The similar mechanism is carried out for holes, and also at the presence of both types of quasi-particles). The concentration of antibonding quasi-particles is critical, when the chain of replacement of a bonding orbital by an antibonding one covers whole crystal (or its macroscopical area) simultaneously and the crystal assimilates to one huge molecule with all weakened bonds. Thus λ_D (T_{mel}) is a correlation length in a task of collective weakening of bonds in a crystal.

As for the criterion of "photomelting" for the case of defect annealing it is necessary to note the following: a physical parameter, on which our hypothesis is based, is the de-Brogle wavelength of antibonding quasi-particles corresponding to temperature of melting of the semiconductor. According to the definition $\lambda_D = h/p$. It is required the quasi-momentum p to be a good quantum number. After according to the hypothesis, λ_D (T_{mel}), and corresponding volume V_D (T_{mel}) (see formula (1)) are fixed, as formally designated length and volume. As a result critical concentration contains parameter $n_{cr} \sim (mT_{mel})^{3/2}$, where m - effective mass of conductivity of quasi-particle. While λ_D (T_{mel}) or m and T_{mel} have their sense, formulas (2) - (4) for n_{cr} work so much as a concept of quasi-momentum, as a "good" quantum number for crystal with defects (or even amorphously macroscopical areas) is maintained. Everything depends on a destruction degree of a basic matrix of a crystal. If de-Brogle wavelength of the particle, $\lambda_D(x) = h/p(x)$ is a smooth function of the coordinate, i.e. the condition of a quasi-classical state $|d\lambda/dx|$ $\ll 1$ is satisfied, then formulas (2) - (4) for n_{cr} are approximately right.

As known, in amorphous semiconductors, generally, quasi-momentum is not a "good" quantum number, there is no Bloch waves, but nevertheless, despite lacking the distant order, a band scheme (the "expanded" states, the forbidden zone, the forbidden zone of mobility, concepts of electron and holes as quasi-particles, exciton, etc.) is maintained. For an example we shall consider Ge and Si. It is known, that in both crystal and amorphous states their basic structural element is tetrahedron with four atoms in bonds. The basic difference between a crystal and an amorphous form is that in the amorphous phase tetrahedrons are oriented to each other in a random way. Thus, it is very important, that the near order is saved, and the breaking of chemical bond does not depend on the distant order. Breaking of chemical bond means releasing of an electron from the bond, and the occurrence of a hole which can participate in conductivity as a quasi-particle. Therefore, if the semiconductor is a material in which for breaking of the bond is required the energy $\Delta E\sim$ (1÷2) eV, the same definition can be used for amorphous semiconductors. Melting means the breaking of a tetrahedral structure and the occurring of a denser packaging. Hence, from a position of chemical bonds the concept of «melting temperature" and quasi-particle with a corresponding de-Brogle wavelength, is saved. So, for example, in amorphous semiconductors the phenomena often take place there are, in which electrons with $kL\leq 1$ i.e. $\lambda_D\geq L$ (k - a wave vector, L - free length of electron) participate [23]. In such cases formulas (1) and (2) for n_{cr} have to be applied, as an estimation.

Proceeding from the above-mentioned facts it is possible to draw a conclusion, that a degree of really achievable photoionization in conditions of experiments at PLA, which for Si and GaAs $n\leq 10^{20}$ sm^{-3} [22], can provide realization of annealing processes without high-temperature heating of a crystal. In this case the experimental results not finding explanations within the thermo-annealing model at PLA [4-6,8-10] become clear.

References

[1] Z.V. Jibuti, N.D.Dolidze, G.Eristavi. // *Technical Physics*, Vol. 53, No. 6, 808-810 (2008).

[2] A.Bibilashvili, N.Dolidze, Z.Jibuti, R.Melkadze, G.Eristavi. // *J.Nanotechnology Perceptions*, 4, 29, N25B107A. (2008).

[3] Z.V.Jibuti, N.D.Dolidze, B.E.Tsekvava. // *Proc. of International scientific conference "Physical and chemical bases of formation and updating micro- and nanostructures*", Kharkov, Ukraine, 10-12 0ctober, 2007, 137-139.

[4] I.G.Gverdciteli, A.B.Gerasimov, Z.V.Jibuti, M.G.Pkhakadze. // *Poverkhnost.* в. 11, 132, (1985).

[5] R.Baltrameyunas, R.Gashka *at al.* // *Phys.Tekh.Polupr*, 21, 2219, (1987).

[6] K.V.Rudenko, S.V.Zhuk, G.G.Gromov. // *Phys.Tekh.Polupr*, 21, 1750, (1987).

[7] Z. V. Dzhibuti, N. D. Dolidze, G. Sh. Narsiya, G. L. Eristavi. // *Tech. Phys. Lett.,* 23, 10, 746, (1997).

[8] S. A. Avsarkisov, Z. V. Jibuti, N. D. Dolidze, and B. E. Tsekvava. // *Tech. Phys. Lett.*, 32, 3,259, (2006).

[9] G.A.Kachurin, E.B.Nidaev. // *Phys.Tekh.Polupr*, 14, 424, (1980).

[10] N.Dolidze, Z.Jibuti, M.Pkhakadze, N.Sikhuashvili. // *Bulletin of the Georgian Academy of Sciences,* 169, 3, 488, (2004).

[11] A.V.Dvurechenski, G.A.Kachurin, E.B.Nidaev, L.S.Smirnov. *Pulsing annealing in semiconductor materials.* // Moscow, 1982. 208 p.

[12] I.B.Kchaibulin, L.S.Smirnov. // *Phys.Tekh.Polupr*, 19, 569, (1985).

[13] J.A.Van Vechten, A.D.Compaan. // *Sol. St.Commun.*, 39, 8, 867, (1981).

[14] V.V.Kapaev, U.V.Kopaev, S.N.Molotkov. // *Microelectronics*, 12, 499, (1983).

[15] U.V.Kopaev, V.V.Meniailenko, S.N.Molotkov. // *Phys.Tv.Tela*, 27, 11, 3288, (1985).

[16] N.D.Dolidze, Z.V.Jibuti, V.N.Mordkovich, B.E.Tsekvava. // *GEN*, №4, 84-87, 2005.

[17] V.N.Abakumov, O.V.Zelenova, *at al.* // *Pis'ma v JTF*, 8, 1365, (1982).

[18] J.I.Alferov, V.N.Abakumov, U.V.Kovalchuk *at al.* // *Phys.Tekh.Polupr*, 17, 235, (1983).

[19] J.I.Alferov, U.V.Kovalchuk *at al.* // *Pis'ma v JTF*,9, 1373, (1983).

[20] J.I.Alferov, U.V.Kovalchuk *at al.* // *Izv. AN SSSR, ser. phyzicheskaya*, 49, 1069, (1985).

[21] V.Lang. Review of Radiation-Induced Defects in III-V Compounds. *Inst. Phys. Conf.* Ser. № 31, (1977), Chapter 1, p.70-94.

[22] V.N.Abakumov, J.I.Alferov, U.V.Kovalchuk, E.L.Portnoi. // *Phys.Tekh.Polupr*, 17, 12, 2224, (1983.

[23] N.F. Mott, E.A. Davis. Electron Processes in Non-Crystalline Materials. vol. 1 and 2. // *Russ.Tr. "Mir",* Moscow, 1982, 664 p.

In: New Developments in Material Science ISBN 978-1-61668-852-3
Editors: E. Chikoidze and T. Tchelidze

Chapter 7

IDENTIFICATION AND MODEL OF DIVACANCY IN GE AND GAAS

N. D. Dolidze**, Z. V. Jibuti and B. E. Tsekvava
Iv.Javakhishvili Tbilisi State University
13 I.Chavchavadze ave, Tbilisi, Georgia, 0179

ABSTRACT

A model of a divacancy which explains a number of experimental facts in irradiated Ge and GaAs is proposed. The model explains the absence of photoconductivity associated with photoexcitation of infrared absorption bands at 0.44, 0.52 eV (Ge) and 0.80, 1.00 eV (GaAs) respective to divacancies; the difference between the values of activation energy of divacancy energy levels are determined by different methods; the existence of the limiting position of the Fermi level in the forbidden gap in the case of irradiation with large radiation fluxes is revealed.

It was shown in [1] that so-called light-sensitive defects are formed in *p*-type germanium, as well as in *n*-type germanium converted into *p*-Ge as a result of irradiation with accelerated electrons (1-5 MeV). Photosensitivity is manifested only in samples possessing p-type conductivity after irradiation [1-6]. It is shown in [4, 6, 7] that light-sensitive defects are responsible for

* E-mail: nugo42@mail.ru

infrared (IR) absorption bands at 0.44 and 0.52 eV. These defects posses in according to temperature dependent Hall measurements energy levels E_v + 0.16 eV and E_v + 0.08 eV in the band gap, depending on the charge state of a defect [5]. At the same time the electrically active levels E_v + 0.52 eV and E_v + 0.44 eV have not been observed in experiments. It should also be noted that photoconductivity was not observed in the case of absorption of photons with energies 0.52 and 0.44 eV. In [4,7] these defects in germanium were called divacancies.

The defects with the similar properties has been found out in IR absorption spectrum of GaAs irradiated with accelerated electrons (1-5 MeV) [8-14]. It was shown, that these defects depending on their charge condition, were responsible for two IR absorption bands (1.00 and 0.80 eV), and for two electrically active levels (E_c - 0,52 eV and the E_c - 0,72 eV) in the band gap [12]. The further researches [13,14 shoud, that researched defect in GaAs has also other similar properties with the divacancy in Ge – dichroism, spatial orientation in a crystal, etc. The assumption was made, that this defect is mixed divacancy ($V_{Ga} + V_{As}$) [13,14].

The similar situation (the difference between the energies of the optical absorption bands ~0.69 and ~0.32 eV, and the energies of the electrically active levels E_c -0.39 eV and E_c- 0.54 eV, respectively [15]) is observed for a divacancy in silicon [16-18]. In this case also the absorption of photons corresponding to energies of optical absorption bands does not lead to photoconductivity.

At the same time, it is well known [19, 20] that irradiation of *n*- and p-type germanium, gallium arsenide and silicon samples leads to the decrease of concentration of the majority charge carriers; i.e., the Fermi level is displaced to the middle of the band gap and attains a certain limiting position, which is the same for both conductivity types (the so-called limiting Fermi level), and changes insignificantly upon a further increase of integrated radiation flux. A unified theory explaining all these experimental facts has not been developed yet.

Nowadays, there exists a model of divacancy in silicon [16] (Figure 1). The same model is used for germanium [7] and can be used for gallium arsenide. (As GaAs has the same cubic lattice, only with the difference: there are two different atoms in unit cell).

According to this model, at the initial stage (before Jahn-Teller distortions appear), there are two vacancies at adjacent atomic sites *c* and *c'* (dashed circles in Figure 1) located along a spatial diagonal of the cube and six neighboring atoms *{a, d, a' d', b, b')* with bent bonds *a-d, a'-d',* and *b—b'.*

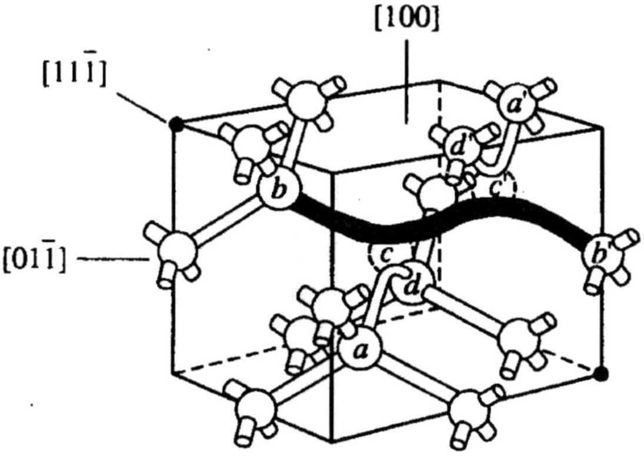

Figure 1. 3D model of a divacancy in silicon (according to [16]).

The corresponding polyatomic nonlinear molecule possesses the D_{3d} symmetry. The group D_{3d} has four one-dimensional and two two-dimensional irreducible representations. Consequently, all electronic terms of the molecule belong to one of the six irreducible representations of the group D_{3d}. Thus, in addition to nondegenerate energy levels corresponding to one-dimensional representations, there also exist doubly degenerate electron energy levels corresponding to two-dimensional irreducible representations (if we disregard spin). A degenerate electron energy level occupied by electrons is unstable to the Jahn-Teller effect. This effect leads to a distortion of the configuration of the nonlinear molecule, and its symmetry is lowered to C_{2h}; the group C_{2h} contains only three one-dimensional irreducible representations [21], and, hence, only nondegenerate electron energy levels are present. The spatial model of a divacancy and its wave functions as given by the LCAO model without and with accouting the Jahn-Teller effect (according to [16]) are presented in Figures 1 and 2. According to this model, a divacancy is a multiply charged center with four energy levels: the two lower levels (*1, 2*) lie extremely close to each other in the immediate vicinity of the top of the valence band, while the other two levels (*5, 4)* lie deep in the band gap. The charge state of the divacancy is determined by the number of electrons executing the *b-b'* bond between the atoms. When this bond is free of electrons (levels *3, 4* in Figure 3), the divacancy is in the state with double positive charge. When one or two electrons (filling level *3)* execute this bond, we have a singly positively charged or neutral divacancy, respectively. This bond can accommodate one or two more electrons at level *4.* In this case, the divacancy acquires a single negative or double negative charge.

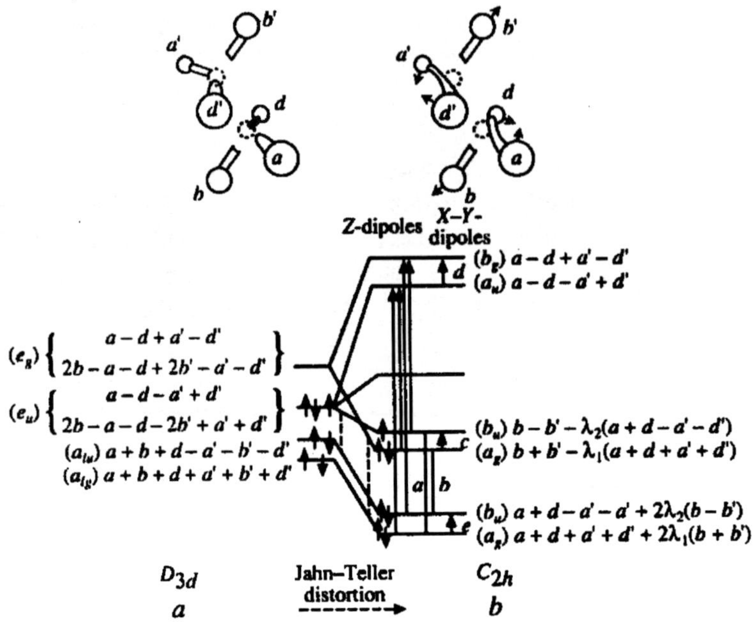

Figure 2. Simple molecular-orbital LCAO model of the electronic structure of a divacancy (a) before Jahn-Teller distortion (D_{3d}) and *(b)* after the distortion (C_{2h}) [16]. Solid arrows denote electrons and their spins for a singly charged positive state, while dashed arrows denote extra electrons for a singly charged negative state.

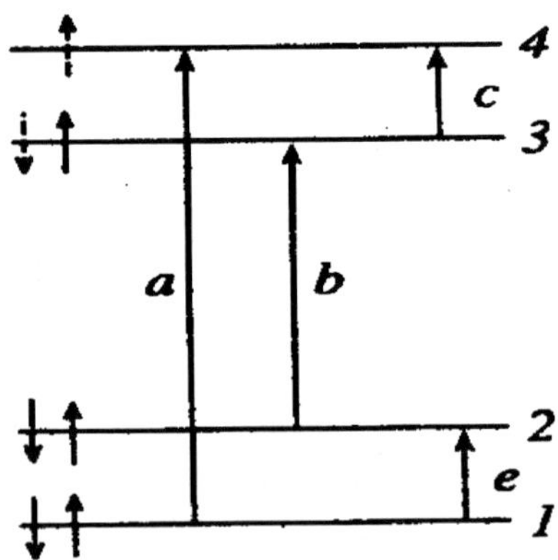

Figure 3. Fragment of Figure 2, showing the levels of a divacancy determining its charge states. Solid arrows denote electrons and their spins for a singly charged positive state, while dashed arrows denote extra electrons for a singly charged negative state.

This model of divacancy can be applied to describe the mixed divacancy in gallium arsenide, as the crystal of gallium arsenide has the same D_{3d} symmetry, with the only difference: from six of broken bonds around divacancy, three belong to atoms of gallium, and others – to atoms of arsenid, one on each atom. And in this case formation of "deformed" coupled (in part covalent, and in part ionic) bond between different type atoms *a-d* and *a'-d'* is expected, and these four electrons being coupled will be placed on orbital *a-d* and *a'-d'*. Others electrons (one for the positive state, three for negative) form bonding *b-b'*.

This model of the divacancy used in [7, 16-18] correctly describes some experimental results but fails to explain all the above-mentioned experimental facts for germanium and silicon.

In order to solve this problem, we propose, in contrast to [7, 16-18], a model of divacancies taking into account the bending of energy bands of a semiconductor by fluctuation electric fields produced by charged defects. We assume that there is no correlation in the distribution of defects in a semiconductor and their concentrations fluctuations are of the Gaussian type [22-24].

Let's apply the method of bent bands to our problem. It was shown in [23] that the fluctuation field of the spatial charge bending the bands is created by chemical impurities. In our case, such impurities are the initial impurities (donors) compensating radiation-induced point defects and their simple complexes, formed in *n*-Ge as a result of irradiation by an integrated flux of accelerated electrons (E = 2-6 MeV, $\Phi = 10^{16}$-$5\text{x}10^{17}$ cm^{-2}) at temperature T = 77 K. This conclusion follows from the fact that conversion of the conduction type (overcompensation) was attained as a result of irradiation of the *n-Ge* samples under investigation with an initial concentration of electrons $n \sim 10^{14}$-10^{16} cm^{-3} and that the samples displayed *p*-type conduction after irradiation. The equilibrium hole concentration in this case did not exceed $\sim 10^{10}$ cm^{-3}. It should be noted that irradiation gives rise to randomly distributed charged centers (defects). In the case of a strong compensation, it is necessary to take into account Gaussian fluctuations of the concentrations of charged impurities and radiation defects, which create a fluctuating large-scale electrostatic potential. The effect of this potential on band bending and on the shift of the Fermi level toward the middle of the band gap must be taken into consideration even for moderate concentrations of defects of 10^{15}-10^{16} cm^{-3}. Under these conditions, the concentration of charged defects creating strong fluctuation fields is high, while screening is weak. However, a low concentration of mobile carriers and a high density of the fluctuation charge give rise to difficulties associated with the taking into account of screening [23-25]. The difficulties associated with screening in the case of a strong compensation were overcome in [23] with the help of the hypothesis that the arrangement of impurities is correlated (for shallow centers), which is determined by the crystal growth technology. In this case, a specific nonelectron mechanism of screening (mutual screening of donors and acceptors) takes place. In the case of radiation defects, such a screening model is not adequate to the actual situation. Since the experimentally determined energy spacing between the lowest and highest position of the top of the bent valence band (potential relief) in germanium is approximately equal to 0.3 eV, the linear-screening approximation not napplicable. The reason for this is that such a high fluctuation potential leads to a deep spatial modulation of the electron density. The linear-screening theory is based on linearization of the Poisson equation, which is justified only in the case of a small depth of the potential well and a weak inhomogeneity of the electron density.

Farther we will now use the basic concepts of the theory of nonlinear screening formulated in [25-27]. In this theory, one of the main difficulties is associated with determining the form of distribution of impurities in a semi-conductor. Following [25-27] we assume that there is no correlation in the distribution of defects, i.e., that the distribution is of the Poisson type and fluctuations are Gaussian (small fluctuations). In order to estimate potential-energy fluctuations, we will use the approximation of a uniformly charged sphere [27]. We assume that fluctuations have the form of homogeneous spherical defect pileups having radius R and characterized by a Poisson distribution of defects in them. We denote the average concentration of defects by N_t. Then, the average number of particles in a sphere is $N = 4\pi N_t R^3/3 = 4N_t R^3$. As known [28] mean-square fluctuation of number of particles for ideal gas enclosed in relatively small volume is equal to $\Delta N = \sqrt{N} = 2\sqrt{N_i R_3}$. A typical charge of a sphere is equal to $2Ze\sqrt{N_t R^3}$ ($Z = \pm 1, \pm 2$), and the mis fluctuation potential energy $y(R)$ of an electron is given by

$$\gamma(R) = 2Ze^2\sqrt{N_t R^3}/\varepsilon R = 2Ze^2\sqrt{N_t R}/\varepsilon \quad (1)$$

where ε is the dielectric constant of the medium and e is the electron charge. As $R \to \infty$, we have $\gamma(R) \to \infty$ in complete accord with the well-known result according to which the rms Coulomb energy fluctuation $\left[N_t \int (e^2/\varepsilon r)^2 dr\right]^{1/2}$ diverges at large distances. This result is physically meaningless, which indicates that screening have to be taken into account in for a random distribution of Coulomb centers. In our case of large-scale fluctuations of randomly distributed charged centers, the only possible mechanism of nonlinear screening that bounds the radius of a sphere, is the electron (hole) screening (in spite of the low concentration of free charge carriers). We introduce the radius of a sphere $R = R_o$ in such a way that the fluctuating charge density $\delta N = 0.5Ze\sqrt{N_t / R_0^3}$ is equal to the hole charge density. Obviously, the holes (electrons) in this volume neutralize the excess density of the negative (positive) charge of the sphere. Consequently, we have

$$p=0.5Z\sqrt{N_t / R_0^3} \text{ or } R_o=\sqrt[3]{Z^2 N_t / 4p^2}\,. \qquad (2)$$

In our experiments, *n*-type germanium is compensated by irradiation: $N_t = N_d + N_a$ (N_d and N_a are the concentrations of radiation defects of the donor and acceptor type, respectively). Considering that compensation has been attained, we can assume that, approximately, $N_t = 2N_d$.

It follows that fluctuations of a size $R > R_o$ are neutralized completely by free carriers, while fluctuations with a size $R < R_0$ remain unscreened. Consequently, an approximate value of y is obtained from Eqs. (1) and (2) to be

$$\gamma = \sqrt[3]{4Z^4 e^6 N_t / \varepsilon^3 p} \qquad (3)$$

For low concentrations of holes, the potential well is found to be quite deep. The electron levels of a divacancy lie in this well (Figure 4). For $p = 10^{10}$ cm $^{-3}$ and $N_d = 10^{15}$ cm^{-3} ($N_t \sim 2N_d$), estimating the amplitude γ of potential-energy oscillations for Ge from Eqs. (3) gives γ=0.16 eV. The condition for the location of the ground-state energy levels in the well, $\hbar^2/2m R_0^2 << \gamma$, [21] can be written, in accordance with Eqs. (2) and (3), in the form $(p/N_t)^{5/3} << a^{-1}N_t^{-1/3}$, where $a = \varepsilon\hbar^2/me^2$ (m is the effective electron mass). Since $p << N_d$, inequality is satisfied with a large margin and energy levels *1* and *2* of the divacancy lie near the bottom of the well (the top of the valence band in the given part of the crystal).

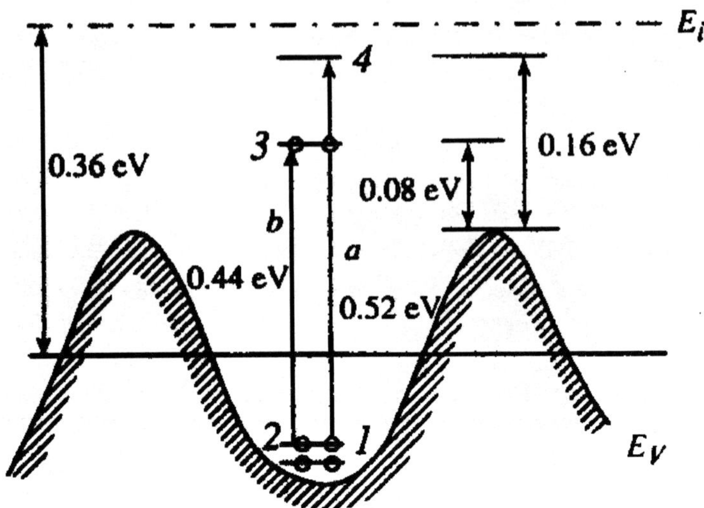

Figure 4. Diagram explaining the difference between optically and electrically determined values of the activation energy of a divacancy in germanium: 0.52 eV is the energy of optical excitation of an electron from the lower level of the divacancy to level E_v + 0.16 eV, while 0.44 eV is the same energy corresponding to level E_v + 0.08 eV (intracenter transitions); E_i is the middle of the forbidden gap of the unexposed crystal. The curve shows the bending of the top of the valence band E_v by the fluctuation potential of charged defects, while the solid straight line corresponds to the top of the unperturbed valence band.

The energy interval between the bottom and the maximum (hump) of the potential relief (bent band) is $\Delta U = 2\gamma \approx 0.32$ eV, while the corresponding experimental value, determined with the help of optical and electrical measurements, is ΔU_{exp} = (0.52 -0.16) eV = (0.44 - 0.08) eV = 0.36 eV (Figure 4). Taking into account that the above values for parameters are typical of our experiments, the agreement between the theoretical model experimental data can be regarded as satisfactory.

For the range of applicability of formulas (2) and (3), it should be noted that the characteristic scales of fluctuations (consequently, the amplitude of oscillations of the potential energy $\gamma(R)$) in a compensated semiconductor with a low concentration of mobile charge carriers in accordance with formulas (2) and (3) can become, indefinitely large (e.g., can exceed the band gap). It was proved in [26], however, that if the bottom of the conduction band is lowered by $0.5E_g$ relative to its position in an unexposed crystal, it will cross the Fermi level in this region of the crystal, intrinsic carriers (electrons) will appear in a number sufficient for screening of the potential well formed, and further lowering of the bottom of the conduction band will stop. Similarly, in the crystal region with an excess of negative charge (potential well for holes), the number of holes formed upon the crossing of the Fermi level will be sufficient for preventing a further rise of the top of the valence band. Consequently, formulas (2) and (3) are inapplicable for the case of low concentrations of charge carriers when $\gamma(R) > 0.5E_g$. Under our experimental conditions, $\gamma(R) < 0.5E_g$, and formulas (2) and (3) hold.

Let's now apply the proposed divacancy model to explain the following experimental facts. Let's consider Ge as an example.

The absence of photoconduction in the case of optical excitation of levels 3 and 4 (transitions a, b in Figure 3). The above estimates indicate that the ground state levels of the divacancy *(1,2)* lie near the bottom of the potential well. When photons with energy 0.44 or 0.52 eV are absorbed, electron transitions from these levels to levels *3* and *4* (intracenter transition) take place. In this case, electrons must perform transitions from the valence band to the lower vacant levels, *1* and *2*; i.e., holes are generated near the bottom of the well in the valence band (consequently, photoconduction must appear). However, the divacancy in the well is in an unstable charge state, and the surplus electron from the upper level instantaneously recombines with a nearest neighbor hole. Therefore, the hole disappears before it can participate in photoconduction.

Difference in the values of activation energy of divacancy levels 3 and 4 (Figure 3) determined by optical and electrical methods. The activation energies of energy levels determined from temperature dependent Hall effect measurement indicate that the Fermi level is shifted towards the middle of the band gap upon an increase of temperature. In this case, electrons can occupy levels 3 and 4; i.e., the divacancy can change its charge state. Electrons perform transitions from the hump of the bent valence band to these levels. The energy required for the excitation of such electrons corresponds to the spacing between the hump of the bent valence band and levels *3* and *4*. Figure 4 shows the diagram explaining the difference between optically and electrically determined values of activation energy of a divacancy in germanium: the optically determined value of the activation energy for level *4* (0.52 eV) corresponds to the electrically active level Ev + 0.16 eV, while the electrically active level Ev + 0.08 eV corresponds to with optical activation energy (0.44 eV) of level 3.

The existence of the limiting position of the Fermi level in the forbidden gap. As mentioned above the existence of the limiting Fermi level in both types of semiconductors was established experimentally. It was shown that the limiting of Fermi level is different for different semiconductors: it lies near the middle of the band gap in silicon and in gallium arsenide, and is closer to the lower edge of the gap in germanium. In n-Ge, a conversion of the conduction type is observed. Different authors give different positions of the limiting Fermi level in the band gap of germanium (from E_y + 0.07 eV to E_v + 0.24 eV) for various temperatures and types of irradiation. In [19], the limiting position of the Fermi level for *p*-Ge and for *n*-Ge converted to the *p* type by irradiation by fast electrons at $T = 77$ K was found to be E_v + 0.07 eV. It was assumed in [19] that an amphoteric level belonging to a complex radiation defect exists in this region. The idea of pinning of the Fermi level near an amphoteric level seems attractive, and we adhere to the same point of view. .However, in contrast to [19], we proved that such an amphoteric center is a multielectron center (divacancy), whose energy level diagram is given in Figure 4. Since $\gamma \sim 0.16$ eV in Ge, according to our estimations, the optically determined levels 0.52 and 0.44 eV must be separated by intervals ~0.16 and 0.08 eV from the hump of the bent valence band, respectively, which corresponds to a distance ~0.1 eV from the middle of the band gap in the unexposed crystal. During irradiation of *n*- and *p*-Ge, divacancies (which are compensating centers in both cases) are introduced. In p-type samples, the divacancy formed can be in the doubly positive (no electrons on level *3*), singly positive (one electron on level *3*), or neutral (two electrons on level *3*) charge states. In the former two cases, the divacancy is a positively charged ionized center and behaves as a donor. Therefore, the Fermi level moves upwards to the middle of the band gap upon irradiation of the sample (i.e., upon an increase of the divacancy concentration). When the Fermi level lies between the level *3* and *4* and is separated from these each of them by at least (2-3)kT, the charge state of the divacancy changes and the divacancy becomes neutral (level *3* is occupied by two electrons, while level *4* is empty). This situation comes about, because the energy interval between level *3* and *4* is $\Delta E > kT$, $\Delta E/kT \geq 10$ ($T = 77$ K). A further upward displacement of the Fermi level as a result of irradiation leads to a transition of the divacancy to the singly or doubly charged state (filling of level *4*), which means that the divacancy (negatively charged ionized center in this case) becomes an acceptor. As a result, the Fermi level starts moving towards the valence band again, and, after it attains the lower level, the divacancy again transforms into a donor, after which the process is repeated. The Fermi level turns out to be fixed between the two levels, which is the reason for the existence of the limiting Fermi level. The same pattern is also observed for *n*-Ge, with the only difference Fermi level moves, as a result of irradiation, toward the middle of the forbidden gap from the conduction band. The similar picture are expected to take place in Si and GaAs, with Fermi level fixed in the middle of band gap where divacancy levels are located.

REFERENCES

[1] A.B. Gerasimov, N. D. Dolidze, N. G. Kakhidze, *et al.* // *Fiz. Tekh. Poluprovodn.* (Leningrad) 1 (7), 982 (1967) [Sov. Phys. Semicond. 1, 822 (1968)].

[2] H. Saito, N. Fukuoka, H. Hattori, and J. H. Crawford, in *Radiation Effects in Semiconductors,* Ed. by F. L. Vook (Plenum, New York, 1968), p. 232.

[3] T. M. Flanagan and E. E. Klontz, *Phys. Rev.* 167, 789 (1968).

[4] H. J. Stein, in *Radiation Damage and Defects in Semiconductors,* Ed. by J. E. Whitehouse (The Institute of Physics, London, 1973), p. 315.

[5] A. R. Basman, A. B. Gerasimov, N. G. Kakhidze, *et al. Fiz. Tekh. Poluprovodn.* (Leningrad) 7 (7), 1347 (1973) [Sov. Phys. Semicond. 7, 903 (1973)].

[6] A. B. Gerasimov, N. D. Dolidze, *et al. Fiz. Tekh. Poluprovodn.* (Leningrad) 11 (7), 1349 (1977) [Sov. Phys. Semicond. 11, 793 (1977)].

[7] A. B. Gerasimov, N. D. Dolidze, R. M. Donina, *et al. Phys. Status* Solidi A 70, 23 (1982).

[8] Brailovskii E.Yu, Brudnyi V.N., *et al.* // *Fiz. Tekh. Poluprovodn.* (Leningrad) 6 (10), 2075 (1972) [Sov. Phys. Semicond. 11, 793 (1972)].

[9] Arefyev K.P., Brudnyi V.N., *et al.* // *Fiz. Tekh. Poluprovodn.* (Leningrad) 13 (6), 1142(1979) [Sov. Phys. Semicond. 11, 793 (1979)].

[10] Z.V.Jibuti, N.D.Dolidze, *et al.* // *Fiz. Tekh. Poluprovodn.* (Leningrad) 21, 930 (1987) [Sov. Phys. Semicond. 11, 793 (1987)].

[11] Z.V.Jibuti, N.D.Dolidze // *Pis'ma v JTF.* (Leningrad) 17(5), 41 (1991) [Sov. Technical Physics Lett. 17, 41 (1991)].

[12] N.Dolidze, Z.Jibuti, G.Cholokashvili, I.Shiriapov. // *Bulletin of the Georgian Academy of Sciences,* v.162,N1, 2000.

[13] Z. V. Dzhibuti, N.D.Dolidze . // *Technical Physics Lett.*, 26(9), 789 (2000).

[14] Z. V. Dzhibuti, N.D.Dolidze // *Technical Physics Lett.,* 27(12), 1008 (2001).

[15] V. S. Vavilov, N. P. Kekelidze, and L. S. Smirnov, *Effects of Radiation on Semiconductors* (Nauka, Moscow, 1988).

[16] G. D. Watkins and J. W. Corbet, *Phys. Rev.* 138, A543 (1965).

[17] L. J. Cheng, J. C. Corelli, J. W. Corbet, and G. D. Watkins, *Phys. Rev.* 152, 761 (1966).

[18] A. H. Kalma and J. C. Corelli, *Phys. Rev.* 173, 734 (1968).

[19] A. B. Gerasimov, *Fiz. Tekh. Poluprovodn.* (Leningrad) 12 (6), 1194 (1978) [Sov. Phys. Semicond. 12, 709 (1978)].

[20] V. S. Vavilov, *Effects of Radiation on Semiconductors* (Fizmatgiz, Moscow, 1963; Consultants Bureau, New York, 1965).

[21] L. D. Landau and E. M. Lifshitz, *Course of Theoretical Physics,* Vol. 3: *Quantum Mechanics: Non-Relativistic Theory* (Nauka, Moscow, 1974; Pergamon, New York, 1977).

[22] V. L. Bonch-Bruevich, *Fiz. Tverd. Tela* (Leningrad) 4 (9), 2660 (1962) [Sov. Phys. Solid State 4, 1953 (1962)].

[23] L. V. Keldysh and G. P. Proshko, *Fiz. Tverd. Tela* (Leningrad) 5 (12), 3378 (1963) [Sov.Phys. Solid State 5, 2481 (1963)].

[24] E. O. Kane, *Phys. Rev.* 131, 79 (1963).

[25] B. 1. Shklovskii and A. L. Efros, *Zh. Eksp. Teor. Fiz.* 60 (2), 867 (1971) [Sov. Phys. JETP 33, 468 (1971)].

[26] B. I. Shklovskii and A. L. Efros, *Zh. Eksp. Teor. Fiz.* 62 (3), 1156 (1972) [Sov. Phys. JETP 35, 610 (1972)].

[27] A. L. Efros, *Usp. Fiz. Nauk* 111 (3), 451 (1973) [Sov.Phys. Usp. 16,789(1973)].

[28] L.D.Landau and E. M. Lifshitz. *Statistical Physics*, Nauka, Moscow, 1964, 568 p (Russian).

In: New Developments in Material Science ISBN 978-1-61668-852-3
Editors: E. Chikoidze and T. Tchelidze

Chapter 8

NANOTUBULAR BORON: GROUND-STATE ESTIMATES

Levan Chkhartishvili*

Department of Physics, Georgian Technical University
77 Merab Kostava Avenue, Tbilisi 0175, Georgia

ABSTRACT

Variety of boron crystalline modifications and boron-rich solids are known to be constructed from the interconnected boron icosahedra. The same short-range structural order is characteristic for amorphous boron. Then, most of B atoms usually are members of the almost regular atomic triangles. This circumstance leads to possibility of the new kind of elemental boron nanostructures in form of flat or rolled boron surfaces with triangular two-dimensional lattices.

Based on the given structural description and using appropriate B−B interatomic potential it is possible to calculate easily their ground-state parameters, as well as the molar binding energy, equilibrium bond length, zero-point vibration frequency etc.

Thereto the quasi-classical B−B potential in dependence on interatomic distance suggested earlier seems to be useful because part of its parameters, like the coefficients of harmonic and anharmonic terms (when the potential-function in the vicinity of equilibrium interatomic distance is approximated by the cubic expression) and atomic core charge

* E-mail: chkharti2003@yahoo.com

(which is near the boron atom valence of 3) had been successfully applied to interpret isotopic composition effect on boron structural and melting parameters, respectively.

There are estimated set of quasi-classical ground-state parameters for the stable boron nanotubes. In particular, corresponding molar binding energy is found to be ~ 8.40 eV / atom.

BORON NANOSTRUCTURES

As it is well known [1], variety of boron crystalline modifications and boron-rich solids mainly are constructed from the interconnected boron icosahedra B_{12} (Figure 1). The same short-range structure is characteristic for amorphous boron [2]. Then, most of B atoms usually are members of the almost regular atomic triangles. This circumstance leads to possibility of the new kind of elemental boron nanostructures in the form of planar or rolled atomic surfaces with triangular two-dimensional lattices (Figures 2-4).

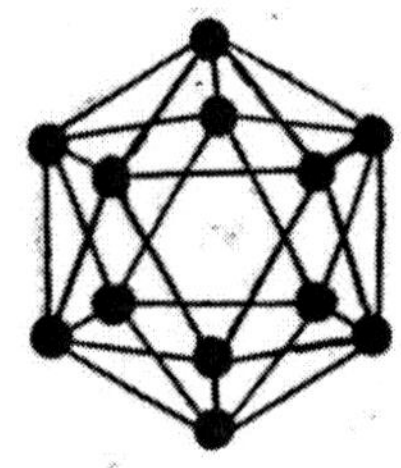

Figure 1. Boron icosahedron B_{12}.

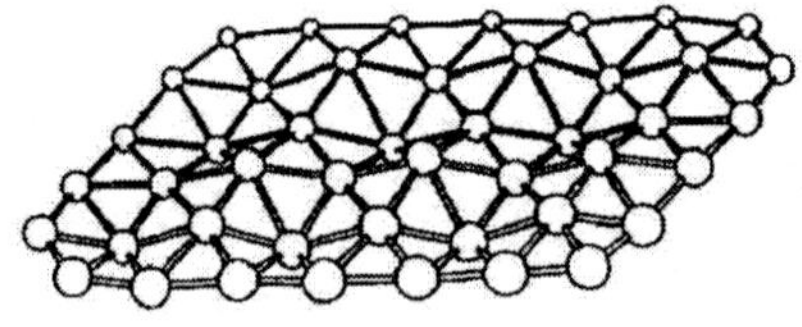

Figure 2. Fragment of the boron quasi-planar (buckled) sheet.

Such geometries for finite B_n clusters experimentally can be estimated based on measurements of apparent ionization potentials and fragmentation patterns for collision-induced dissociation [3].

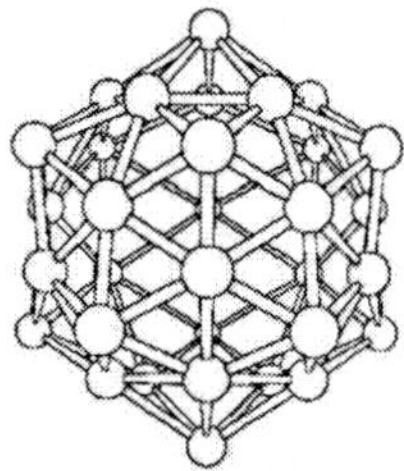

Figure 3. Boron fullerene.

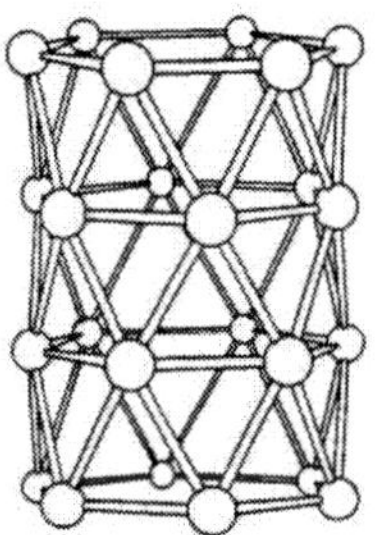

Figure 4. Boron nanotube.

Theoretically nanostructural formation of boron clusters was first proposed by Boustani [4]. According to local-spin-density calculations (including nonlocal corrections), most of structures prefer planar or quasi-planar arrangements (the latter is generated from planar one by forcing the boron atoms out of plane, up and down, and in periodic fashion) and can be considered as fragments of a surface or as segments of a sphere.

Experimentally the atomic and electronic structures and chemical bonding of small boron clusters up to 20 atoms were probed by the photoelectron spectroscopy combined with appropriate quantum mechanical calculations. In particular, clusters B_5^- and B_5 were investigated [5] using anion photoelectron spectroscopy. The global minimum of B_5^- was found to have a planar structure with a closed-shell ground-state. Excellent agreement between ab initio detachment energies and experimental spectra was observed, firmly establishing the ground-state-structures for both B_5^- and B_5. Hepta- and octa-coordinated boron atoms in molecular wheels B_8 and B_9 were observed and confirmed in [6]. It was also reported [7] about experimental and theoretical evidences that small boron clusters prefer planar structures and exhibit aromaticity and antiaromaticity according to the Hückel rules. Aromatic boron clusters posses more circular shapes, whereas antiaromatic ones are elongated. Thus, planar boron clusters can be considered as only series of molecules representing a new dimension of boron chemistry. Most of the small boron clusters B_n^+, B_n, and B_n^- with $n = 2 - 15$ as individual species in the gas phase have been reviewed in [8]. It was emphasized that free small boron clusters were characterized using photoelectron spectroscopy and ab initio calculations, which established their planar or quasi-planar shapes. Experimentally quasi-planarity of the small boron clusters also were confirmed in [9,10].

Experimental investigations and computational simulations revealed [11] that boron clusters, which favour planar structures up to 18 atoms, prefer three-dimensional structures beginning at 20 atoms. Neutral cluster B_{20} was found to have a double-ring tubular structure, even though the B_{20}^- remains planar. As for the anion B_{20}^- itself, the tubular structure was shown to be isoenergetic to two-dimensional structures, which were observed and confirmed by photoelectron spectroscopy. The two-dimensional – three-dimensional transition observed at B_{20} suggests that it may be considered as the embryo of the thinnest single-walled all-boron nanotubes. In [12] it was reported first synthesis of pure boron single-walled nanotubes by a chemical reaction over a catalyst with parallel, uniform-radius cylindrical pores. According to recent communication, the multi-walled boron nanotubes also have been synthesized (see reference in [13]).

A systematic theoretical study on B_n clusters carried out applying first-principles quantum-chemical methods to determine their geometric structures show that they are composed of two fundamental units: either of hexagonal or pentagonal pyramids [14]. The resulting quasi-planar and convex structures can be considered as fragments of planar surfaces and segments of nanotubes or hollow spheres, respectively [15,16]. According to the Hartree–Fock self-consistent-field geometry optimization performed in [17] using Slater-type orbitals basis set, fragment of the quasi-planar boron structure is more stable than its isomers in form of isolated unit cell of the real α-rhombohedral boron crystal or predicted boron quasi-crystals. The nanotubular isomers composed of parallel-lying (staggered) atomic rings in turn are shown to be even more stable. As it was mentioned, both predicted quasi-planarity of boron clusters [18] and boron tubular forms [19] were confirmed experimentally. Recently, density-functional-theory has been employed [20] to calculate the electronic and geometric structures, total and binding energies, harmonic frequencies, point symmetries, and highest-occupied-molecular-orbital–lowest-unoccupied-molecular-orbital gaps of small boron clusters $B_n (n = 2 - 12)$. The linear, planar, convex, quasi-planar, cage, and open-cage structures were found. The first and second energy differences have been used to obtain most stable sizes of clusters.

None of the lowest-energy-structures and their isomers is shown to have an inner atom. Later, the studies have been extended on B_n clusters with $n=13-20$ (see [21]). They are found to form additional diverse structural motifs, which ultimately can yield distinct nanostructures. Within whole range of $n=2-20$, almost all physical properties of B_n clusters are size-dependent. In particular, vibration frequencies increase with size. The double-ring tubular form of B_{20} cluster has found to have the highest binding energy per atom.

First-principles calculations reported in [22] also reveal that boron can form wide variety of metastable planar and tubular forms. The preferred planar structure is a buckled triangular lattice that breaks the threefold ground-state degeneracy of the flat triangular plane. When the plane is rolled into a tube, this degeneracy leads to a strong chirality dependence of the binding energy.

The achiral $(n,0)$-tubes derive their structure from the flat triangular plane, while the achiral (n,n)-tubes arise from the buckling plane, and have large cohesion energies and different structure as a result.

Density functional theory calculations carried out [23] to obtain geometric structure of boron sheets and nanotubes show that buckled boron sheet with certain buckling height is more stable than the flat sheet. But, boron nanotubes formed by rolling up of a boron sheet exhibit buckling surfaces if helicity allows for the formation of alternating up and down B-atoms rows in the surface. In all other cases, boron nanotubes exhibit only flat surfaces. Hence, all the $(n,0)$-tubes have to prefer buckled geometry, while not all (n,n)-tubes do the same.

For analysis of the boron nanotubular structures and future purposeful designing devices based on nanotubular boron it is important to predict reliably the sizes of nanotubes with given indexes. Here this task is solved for the mostly regular forms with equal B−B lengths, which exhibit flat surfaces. Namely, the quasi-classical B−B pair interatomic binding energy curve is constructed, and then the expressions of radii of the zigzag and armchair boron nanotubes in terms of B−B bond length are obtained considering geometries of such structures.

Quasi-Classical B–B Potential

Previously, it had been demonstrated [24,25] that an atom can be considered as a quasi-classical electron system in sense of Maslov criterion and, on this basis, the quasi-classical approach to the construction of interatomic pair potentials had been elaborated. Within the initial quasi-classical approximation, using quadratic polynomial approximation for the electron screening factor standing in the effective atomic potential the binding energy $E = E(d)$ between pair of the identical (e.g. boron) atoms was expressed as

$$-4E(d) = \sum_{i,j=0}^{i,j=s} (n_i U_j + n_j U_i)(W(R_i, R_j, d) + W(R_{i+1}, R_{j+1}, d) - W(R_i, R_{j+1}, d) - W(R_{i+1}, R_j, d)) +$$

$$+ 2A_0 \sum_{i=0}^{i=s} (A_i / d + B_i + C_i d)\theta(d - R_i)\theta(R_{i+1} - d).$$

Here d is the interatomic distance; n_i, U_i, A_i, B_i, and C_i are the known quasi-classical parameters of the electron-density and potential distributions in the interacting atoms; R_i is the effective radius of electron cloud affecting an electron of the i th subshell, while s is the total number of subshells in an atom. $W(R_i, R_j, d)$ is a known algebraic function expressing dependence of two spheres' intersection region volume upon their radii R_i and R_j, and centre-to-centre distance d; and $\theta(R_i - d)$ is a Heaviside step-function. These parameters and electron cloud radii can be obtained by fitting quasi-classical electron energy levels and mean orbital radii of the electrons to their first-principles, e.g. Hartree–Fock, values.

Such quasi-classical B–B binding energy $E = E(d)$ in dependence on interatomic distance d seems to be useful for numerical estimations. Part of its parameters, like the coefficients of harmonic and anharmonic terms $c \approx 0.12$ au and $g \approx 4.0$ au in the corresponding interatomic potential $U(d)$ which in the vicinity of equilibrium distance $d = d_0$ is approximated by the cubic function $U(d_0) + c(d - d_0)^2 - g(d - d_0)^3$, and atomic core charge of $A_0 \approx 3.151$ (which is near the boron atom valence of 3) were successfully applied to interpret isotopic composition effect on boron structural [26] and melting parameters [27], respectively. Analysis of the binding energy curve (converted from potential energy one using virial theorem for system total energy) based on the parameters obtained equating quasi-classical electron energy levels and mean orbital radii of the electrons with their Hartree–Fock values leads to the equilibrium B−B distance of 3.37 au (1.78 Å), binding energy at that distance of 0.103 au (2.80 eV), and relative vibration frequency of 0.00480 au (0.131 eV). These values slightly, only by few percents, are deviated from the averaged experimental B−B bond length in main structural units of boron and boron-rich compounds – icosahedra of 3.40 au (1.80 Å), dissociation energy of 0.099 au (2.69 eV), and oscillatory quantum of 0.00479 au (0.130 eV) of diboron molecule B_2, respectively. The binding energy curve obtained in initial quasi-classical approximation for interacting boron atoms is expressed by a piece-wise analytical function. Evidently, for numerical calculations the tabulated form would be more appropriate. Here such form (in atomic units) is given (Table 1) over the range from $d = 3.20$ up to 3.50 au (i.e. in the vicinity of equilibrium interatomic distance of 3.37 au) with step of argument of 0.01 au, what seems sufficient to provide accurate energy computations.

BORON NANOTUBES' GEOMETRIES

As for the procedure of the determination of boron nanotubes radii, it is following.

If $d_{(n,0)}$ is the B−B bonds lengths in a zigzag (n,0) nanotube, the height of the equilateral triangle built up from the boron atoms in vertexes equals to $\delta_{(n,0)} = \sqrt{3}\, d_{(n,0)} / 2$. On the other hand, the regular $2n$-gon with sides of $\delta_{(n,0)}$ is inscribed into the circle with radius coincident the tube radius $r_{(n,0)}$. Consequently, $r_{(n,0)} = \delta_{(n,0)} / 2\sin\pi / 2n$ and

$$r_{(n,0)} = \frac{\sqrt{3}\, d_{(n,0)}}{4\sin\pi / 2n}.$$

Let $d_{(n,n)}$ and $r_{(n,n)}$, respectively, denote the B−B bond length and radius in an armchair (n,n) nanotube. It is clear that the regular n-gon with sides of $d_{(n,n)}$ is inscribed into the circle with radius of $r_{(n,n)}$. Consequently,

$$r_{(n,n)} = \frac{d_{(n,n)}}{2\sin\pi/n}.$$

Nanotube index $n = 1,2,3,...$ determines the number of atoms because the nanotube unit cell consists of $2n$ boron atoms. At $n = 1$ zigzag nanotube degenerate in zigzag atomic chain, while armchair nanotube degenerate in straight atomic chain. Correspondingly, formula obtained for $r_{(n,n)}$ in general case, does not works; instead it should be assumed $r_{(1,1)} = 0$.

Now about the detailed regular geometries of the zigzag and armchair boron nanotubes, which we will describe using cylindrical coordinates $\vec{r}(\rho,\varphi,z)$, which is useful in tubular binding energy and electronic structure calculations.

Constant of the zigzag (*n*,0) nanotube's one-dimensional lattice equals to $d_{(n,0)}$. Its unit cell consists of 2 equidistant atomic rings in parallel planes perpendicular to the axis, each with n boron atoms. Evidently, cylindrical coordinate ρ for all atomic sites equals to tube radius:

$$\rho = r_{(n,0)}$$

As for the coordinates φ and z, they equal to

$$\varphi = (4k - 3 - (-1)^l)\pi/2n,$$
$$z = l\,d_{(n,0)}/2.$$

Here $k = 1,2,3,\ldots,n$ and $l = 0,\pm1,\pm2,\ldots$ number atomic sites in given plane and atomic planes, respectively.

If $d_{(n,n)}$ is the B–B bond length in an armchair (*n*,*n*) nanotube, the side $\delta_{(n,n)}$ of the regular $2n$-gon inscribed into the circle with tube radius of $r_{(n,n)}$ is determined as $\delta_{(n,n)} = 2r_{(n,n)}\sin\pi/2n$. One-dimensional lattice constant of the armchair nanotube equals to the doubled leg of a right triangle with another leg of $\delta_{(n,n)}$ and hypotenuse of $d_{(n,n)}$, i.e., according to the

Pythagorean theorem, $2\sqrt{d_{(n,n)}^2-\delta_{(n,n)}^2}=\sqrt{4-1/\cos^2\pi/2n}\,d_{(n,n)}$. Its unit cell also consists of 2 atomic rings in parallel planes perpendicular to the tube axis and, from its part, each ring consists of n boron atoms. Coordinate ρ for all atomic sites again equals to tube radius:

$$\rho_{\mathrm{B}}=r_{(n,n)},$$

while the rest cylindrical coordinates equal to

$$\varphi=(4k-3-(-1)^l)\pi/2n,$$
$$z=l\,d_{(n,n)}\sqrt{1-1/4\cos^2\pi/2n},$$

where again $k=1,2,3,\ldots,n$ and $l=0,\pm1,\pm2,\ldots$.

On the basis of above obtained relations, it was found the squared distances $(\vec{r}-\vec{r}_0)^2=4\rho^2\sin^2\varphi/2+z^2$ between arbitrary atomic site $\vec{r}\,(\rho,\varphi,z)$ and site $\vec{r}_0\,(\rho,0,0)$ of the so-called central atom with $l=0$ and $k=1$, i.e. with $\varphi=0$ and $z=0$, in zigzag

$$\frac{({}^{kl}_{(n,0)}\mathrm{B}-{}^{00}_{(n,0)}\mathrm{B})^2}{d_{(n,0)}^2}=\frac{3\sin^2(4k-3-(-1)^l)\pi/4n}{4\sin^2\pi/2n}+\frac{l^2}{4},$$

and armchair tubes

$$\frac{({}^{kl}_{(n,n)}\mathrm{B}-{}^{00}_{(n,n)}\mathrm{B})^2}{d_{(n,n)}^2}=\frac{\sin^2(4k-3-(-1)^l)\pi/4n}{\sin^2\pi/n}+\left(1-\frac{1}{4\cos^2\pi/2n}\right)l^2$$

GROUND-STATE ESTIMATES

Using quasi-classical B−B interatomic pair potential and based on above given description of the boron nanotubes' geometries it is possible to calculate easily their ground state parameters, like the equilibrium bond length, equilibrium molar binding energy per, zero-point vibration frequencies etc.

Firstly the sizes of small single-walled B-nanotubes were estimated (Table 2) assuming bond lengths as $d_0 \approx 1.78$ Å, the equilibrium interatomic distance value according to the quasi-classical B–B pair potential. Actually, $d_{(n,0)}$ and $d_{(n,n)}$ depend on n; and their exact values can not be determined geometrically.

Table 1. Tabulated boron–boron binding energy function

d, au	$E(d)$, au
3.20	0.101408
3.21	0.101656
3.22	0.101887
3.23	0.102100
3.24	0.102297
3.25	0.102478
3.26	0.102642
3.27	0.102791
3.28	0.102925
3.29	0.103043
3.30	0.103147
3.31	0.103236
3.32	0.103311
3.33	0.103372
3.34	0.103420
3.35	0.103454
3.36	0.103475
3.37	0.103483
3.38	0.103479
3.39	0.103462
3.40	0.103433
3.41	0.103393
3.42	0.103341
3.43	0.103278
3.44	0.103203
3.45	0.103118
3.46	0.103023
3.47	0.102917
3.48	0.102802
3.49	0.102677
3.50	0.102542

Table 2. Estimated radii of small-sized boron zigzag and armchair nanotubes

$(n,0)$ or (n,n)	$r_{(n,0)}$ or $r_{(n,n)}$, Å
(1,1)	0.00
(1,0)	0.77
(2,2)	0.89
(3,3)	1.03
(2,0)	1.09
(4,4)	1.26
(5,5)	1.51
(3,0)	1.54
(6,6)	1.78
(4,0)	2.01
(7,7)	2.05
(8,8)	2.33
(5,0)	2.49
(9,9)	2.60
(10,10)	2.88
(6,0)	2.98
(11,11)	3.16
(12,12)	3.44
(7,0)	3.46

In general, this task requires solving of the physical problem of the binding energy maximization for given nanotubular structure with respect B−B bond length. But, these dependences evidently have to be weak and in all cases in satisfactory approximation $d_{(n,0)}$ and $d_{(n,n)}$ can be substituted by the mentioned value of 1.78 Å.

Analogously, one can estimate molar binding energy of a boron nanotube. On the one hand, any constituent boron atom has 6 nearest neighbours. On the other hand, binding energy between pair of boron atoms is about 2.80 eV. So, taking into account interaction only between nearest neighbours the tube binding energy per atom may be calculated as 2.80 eV × 6 / 2 = 8.40 eV. As for the zero-point vibration frequency, its value can be found within the same approximation taking into account that frequency of the one-dimensional relative B−B vibrations is 0.131 eV, and boron atom localized near a site on tubular surface participates in 3 independent motions: 0.131 eV × $\sqrt{6/3}$ ≈ 0.185 eV. Of course, these are only rough estimations: actually bond length, molar binding energy and vibration quantum should depend on the tube index.

The proposed quasi-classical B−B binding energy curve together with obtained explicit formulas describing tubular geometries should be useful for precise calculations of the boron nanotubes' ground-state parameters. By them the absolute and relative stabilities of regular boron nanotubes can be determined. From their part, these results can serve as a relevant basis for further investigations of the boron nanotubes' electron structures.

References

[1] Tsagareishvili, G. V.; Antadze, M. E.; Tavadze, F. N. *Production and Structure of Boron* 1991, Tbilisi, Metsniereba.

[2] Volkov, V. V.; Yur'ev, G. S.; Myakishev, K. G.; Il'inchik, E. A. *Vacuum Nanotechnology and Equipment* 2006, 1, Kharkiv, KhPTI, 336-347.

[3] Hanley, L.; Whitten, J. L.; Anderson, S. L. *J. Phys Chem.* 1988, 92, 5803-5812.

[4] Boustani, I. J. *Quant. Chem.* 1994, 52, 1081-1111.

[5] Zhai, H.-J.; Wang, L.-Sh.; Alexandrova, A. N.; Boldyrev, A. I. *J. Chem. Phys.* 2002, 117, 7917.

[6] Zhai, H.-J.; Alexandrova, A. N.; Birch, K. A.; Boldyrev, A. I.; Wang, L.-*Sh. Angew. Chem. Int. Ed.* 2003, 42, 6004-6008.

[7] Zhai, H. J.; Kiran, B.; Li, J.; Wang, L.-Sh. Nat. Mat. 2003, 2, 827-833.

[8] Alexandrova, A. N.; Boldyrev, A. I.; Zhai, H.-J.; Wang, L.-*Sh. Coord. Chem.* 2006, 250, 2811-2866.

[9] Perkins, C. L.; Trenary, M.; Tanaka, T. *Phys. Rev.* B 1998, 58, 9980-9989.

[10] Cao, L. M.; Sun, L. L.; Gao, C. X.; He, M.; Wang, Y. Q.; Li, Y. C.; Zhang, X. Y.; Li, G.; Zhang, J.; Wang, W. K. *Adv. Mater.* 2001, 13, 1701-1704.

[11] Kiran, B.; Bulusu, S.; Zhai, H.-J.; Yoo, S.; Zeng, X. Ch.; Wang, L.-Sh. *Proc. Natl. Acad. Sci.* (USA) 2005, 102, 961-964.

[12] Ciuparu, D., Klie, R. F.; Zhu, Y.; Pfefferle, L. *J. Phys. Chem.* B 2004, 108, 3967-3969.

[13] Mukhopadhyay, S.; Pandey, R.; Yap, Y. K.; Boustani, I. *16th Int. Symp. Boron, Borides and Rel. Mat.* 2008, Matsue, KM, 62.

[14] Boustani, I. *Phys. Rev.* B-II 1997, 55, 16426-16438.

[15] Boustani, I. J. *Solid State Chem.* 1997, 133, 182-189.

[16] Boustani, I.; Quandt,. A. *Euro Phys. Lett.* 1997, 39, 527-532.

[17] Boustani, I.; Quandt, A.; Rubio, A. *J. Solid State Chem.* 2000, 154, 269-274.

[18] Boustani, I. *Surf. Sci.* 1997, 370, 355-363.

[19] Boustani, I.; Quandt, A.; Hernández, E.; Rubio, A. *J. Chem. Phys.* 1999, 110, 3176.

[20] Atiş, M.; Özdoğan, C.; Güvenç, Z. B. *Int. J. Quan. Chem.* 2007, 107, 729-744.

[21] Atiş, M.; Boyukata, M,; Özdoğan, C.; Yildirim, E.; Güvenç, Z. B. 2008, 16th Int. Symp. Boron, Borides and Rel. Mat., Matsue, KM, 33.

[22] Evans, M. H.; Joannopoulos, J. D.; Pantelides, S. T. *Phys. Rev.* B 2005, 72, 045434.

[23] Cabria, I.; Alonso, J. A.; López, M. J. *Phys. Status Solidi* A 2006, 203, 1105-1110.

[24] Chkhartishvili, L.; Lezhava, D.; Tsagareishvili, O.; Gulua, D. Trans. *AMIAG* 1999, 1, 295-300.

[25] Chkhartishvili, L.; Lezhava, D.; Tsagareishvili, O. *J. Solid State Chem.* 2000, 154, 148-152.

[26] Chkhartishvili, L. S.; Gabunia, D. L.; Tsagareishvili, O. A. *Powd. Metal. and Met. Cer.* 2008, 47, 616-621.

[27] Chkhartishvili, L. S.; Gabunia, D. L.; Tsagareishvili, O. A. *Inorg. Mater.* 2007, 43, 594-596.

In: New Developments in Material Science ISBN 978-1-61668-852-3
Editors: E. Chikoidze and T. Tchelidze

Chapter 9

DESCRIPTION OF MAGNETIZATION OF HIGH-TEMPERATURE CERAMIC $YBa_2Cu_3O_{7-x}$ SUPERCONDUCTOR WITH ACCOUNTING OF POLYCRYSTALLINE STRUCTURE OF THE SAMPLE

K. Gamkrelidze, M. Mirzoeva, G. Shonia and G. Gamtsemlidze*

Dep. of Physics, Tbilisi State University,
Tbilisi, 0128, Georgia

ABSTRACT

We present the measurements and analytical expressions for the magnetization at T=77 K of the high-temperature ceramic superconductor $YBa_2Cu_3O_{7-x}$. The behaviour of magnetization is studied as a function of external magnetic field. The sample was prepared by the method of solid-state reaction. Magnetization measurements were conducted with ballistic method. Analytical expressions are obtained in Bean's critical state model by taking into account the field H_{c1}. Description of experimental hysteresis of the polycrystal structure of the sample and granule's anisotropy are taken into account as well.

* E-mail: ketevan_gamkrelidze_07@yahoo.com

In the present work for the description of magnetization hysteresis of high-temperature ceramic $YBa_2\ Cu_3\ O_{7-x}$ superconductor the structure of a sample is taken into account.

The sample was prepared by the method of solid-state reaction. It has cylinder form with diameter 15 mm, height 20 mm, demagnetization factor 0.007, critical temperature (92±1) K. Magnetization measurements were conducted with ballistic method by means of change determination, passing through galvanometer chain while transferring the sample from one measuring coil into another, being connected to meet each other [1].

CALCULATIONS AND RESULTS

It is considered as a polycrystal consisting of monocrystals (granules) oriented arbitrarily relative to the external magnetic field H_0.

Average magnetization of $\overline{M}$ polycrystal is determined by:

$$\overline{M} = \int_0^{\pi/2} M(\gamma)\sin\gamma d\gamma \tag{1}$$

where M (γ) is the magnetization of granules oriented at $\gamma = \vec{C}\,^\wedge \vec{H}$ angle ($\vec{C}$ is the principle axis of a lattice) with respect to magnetic field and sin γdγ is the hit probability of granules in (γ, γ+dγ) interval. According to [2] the dependence of the first H_{c1} critical magnetic field of a granule at γ angle has the form

$$H_{c1}(\gamma) = H_{c1}^c (1 + \chi \sin^2 \gamma)^{-1/2} \tag{2}$$

where H_{c1}^c is the first critical field of the granules oriented along the external magnetic field χ is the anisotropy parameter. For the $YBa_2\ Cu_3\ O_{7-x}$ sample $\chi \approx 25$ [3]. The equation of critical state for the monocrystal oriented at γ angle with respect to H_0 field in axisimmetric case [4] has the form:

$$\frac{dB}{dr} = \frac{4\pi}{c} j_c \tag{3}$$

where B is the local magnetic induction, j_c is the critical current density.

For the boundary conditions we have induction equality to equilibrium value at the boundary

$$B(r) = B_{eq}(H_0) \tag{4}$$

To determine $B_{eq}(H_0)$ dependence, an approximation is used [5]

$$\begin{cases} B_{eq}(H_0) = 0, & H_0 < H_{c1}(\gamma) \\ B_{eq}(H_0) = H_0 - H_s, & H_0 > H_{c1}(\gamma) \end{cases} \tag{5}$$

where

$$H_s \equiv H_{c1} - H_b \tag{6}$$

H_b is the parameter depending on the first critical field

$$H_b = aH_{c1}(\gamma), \qquad 0 < a < 1 \tag{7}$$

The boundary condition (4), with account of (5) takes the form:

$$B(R) = H_0 - H_s \tag{8}$$

The solution of (3) by boundary condition of (8) gives:

$$B(r) = H_0 - H_s + (4\pi / c) j_c (r - R) \tag{9}$$

Magnetization of the .granules oriented at γ angle equals to:

$$4\pi M(\gamma) = \langle B(\gamma) \rangle - H_0 \tag{10}$$

The determine the polycrystal average magnetization $\overline{M}$, account of expressions $4\pi M(\gamma)$ for different values of external magnetic field in (1) gives the following equations for increasing fields:

$$4\pi\overline{M} = K\left\{-\frac{1}{3}\Omega - aH_{c1}^{ab}\mathrm{X}\arcsin\frac{1}{\mathrm{X}} - H_0\right\}$$

$$0 \le H_0 \le H_{c1}^{ab} \tag{11}$$

Here and bellow

$$\left[1-\frac{1}{\chi}(1+\chi)\left(\frac{H_{c1}^{ab}}{H_0}\right)^2 -1\right]^{1/2} \equiv g(x)\ ,\ \left(\frac{\chi+1}{\chi}\right)^{1/2} \equiv \mathrm{X},\ \frac{4\pi}{c} j_c R \equiv \Omega$$

$$\left[\frac{1}{\chi}(1+\chi)\left(\frac{H_{c1}^{ab}}{H_0}\right)^2 -1\right]^{1/2} \equiv g_1(x).$$

$$4\pi\overline{M} = K\left\{\left(-\frac{1}{3}\Omega - H_0\right)(1-g(x)) + \frac{1}{3}\Omega\, g(x) - (1-a)H_{c1}^{ab}\mathrm{X}\arcsin\left(\frac{1}{\mathrm{X}}\cos\arcsin g_1(x)\right) - \right.$$
$$\left. - aH_{c1}^{ab}\mathrm{X}\left\{\arcsin\frac{1}{\mathrm{X}} - \arcsin\left[\frac{1}{\mathrm{X}}\cos\arcsin g_1(x)\right]\right\}\right.$$

$$H_{c1}^{ab} \le H_0 \le H_{c1}^{c} \tag{12}$$

$$4\pi\overline{M} = K\left\{-\frac{1}{3}\Omega - (1-a)H_{c1}^{ab}\mathrm{X}\arcsin\frac{1}{\mathrm{X}}\right\}$$

$$H_{c1}^{c} \le H_0 \le H_{0\,\max} \tag{13}$$

For decreasing fields:

$$4\pi\overline{M} = K\left(1-\frac{H_{0\max}-H_0}{2\Omega}\right)\left(\frac{1}{3}H_{0\max} - \frac{2}{3}\Omega - \frac{1}{3}H_0\right) - K\left\{(1-a)H_{c1}^{ab}\mathrm{X}\arcsin\frac{1}{\mathrm{X}} + \frac{1}{3}\Omega\right\}$$
$$\widetilde{H}_0 \le H_0 \le H_{0\max} \tag{14}$$

$$4\pi\overline{M} = K\left\{\left(\frac{1}{3}\Omega - H_0\right)(1-g(x)) + \frac{1}{3}\Omega\, g(x) + \right.$$

$$aH_{c1}^{ab}X\left\{\arcsin\frac{1}{X} - \arcsin\left[\frac{1}{X}\cos\arcsin g_1(x)\right]\right\} -$$

$$\left.\begin{array}{l} -(1-a)X\, H_{c1}^{ab} \arcsin\left[\dfrac{1}{X}\cos \arcsin g_1(x)\right] \end{array}\right\} \quad H_{c1}^{ab} \le H_0 < \widetilde{H}_0 \qquad (15)$$

$$4\pi\overline{M} = K\left\{\frac{1}{3}\Omega + aH_{c1}^{ab} X \arcsin\frac{1}{X} - H_0\right\},$$

$$0 \le H_0 \le H_{c1}^{ab} \qquad (16)$$

where

$$\widetilde{H}_0 = H_{0\,\max} - 2\Omega \qquad (17)$$

H_{c1}^{ab} Denotes the first critical field of the granules oriented perpendicularly with respect to the external field, K is the coefficient connected with the imperfect behaviour of Meissner effect and it is determined in the linear region of hysteresis by $4\pi M/H_0$ relation, which, for our sample equals to 0.33. $H_{0\,\max}$=320 Oe and $\frac{4\pi}{c} j_c R = 50$ Oe, $H_{c1}^{ab} = 50$ Oe parameters in equations (11-16) are selected on the basis of experimental data.

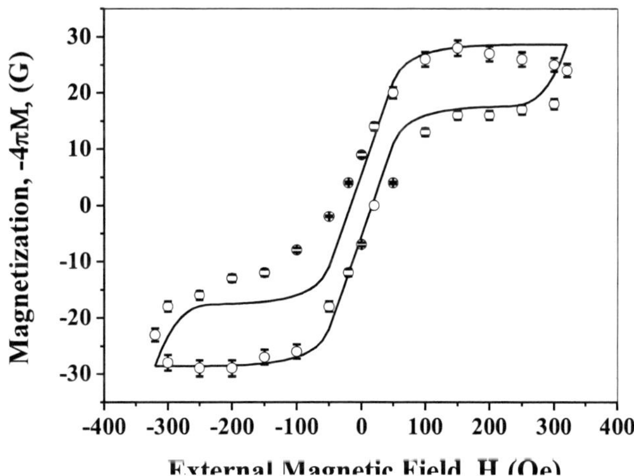

Figure 1. The dependence of magnetization on external magnetic field: symbols – experimental data; solid line – theoretical results.

The figure 1. Presents experimental (symbols) and theoretical (line) results, where the values of a parameter are taken as equal to zero.

Conclusions

Good agreement between theoretical loop and experimental data allows us to conclude that while describing magnetic properties of high-temperature ceramic superconductor obtained by solid state reaction, one should take into account the anisotropy of the granules.

References

[1] G. Gamtsemlidze, , K.Gamkrelidze, M. Mirzoeva, G. Shonia, *Bulletin of The Georgian Akademy of Sciences.* V.164, #3,2001

[2] A.Balacki, L.Burlachkov, L.Gorkov. *Journal of Experimental and Theoretical Physics* (JETP), 90, (1986).

[3] I. Belorusov, B. Gorbunov, V. Smigla,. V Fesenko. *Uspekhi Fizicheskikh Nauk* (UFN),11, (1990).

[4] C Bean. *Rev.Mod.Phys.*36, 31, (1964).

[5] M. McElfresh, Y. Yeshurun, A. Malozemoff, F. Holtzberg, *Physica* A.1,168 (1990)

[6] G. Gamtsemlidze, K. Gamkrelidze, M. Mirzoeva, G. Shonia. *Fizika Nizkikh Temperatur*, (FNT), 19, 9(1993).

In: New Developments in Material Science ISBN 978-1-61668-852-3
Editors: E. Chikoidze and T. Tchelidze

Chapter 10

CARBON IN SIGE ALLOYS BULK CRYSTALS

E. Khutsishvili, L. Gabrichidze and N. Kobulashvili
F.Tavadze Institute of Metallurgy and Materials Science of Ministry of Education and Science, Tbilisi

ABSTRACT

Carbon is usually a foreign unforeseen impurity in semiconductor materials in contrast to elements of III and V group of Periodic System. But carbon is also widely used in construction of modern devices. Therefore, investigation of carbon behavior, as an impurity, especially, in SiGe alloys is very relevant. Si-Ge alloys attract considerable interest in advanced thermoelectric, electronic and optoelectronic applications. There are few data in literature under this question. Si and SiGe alloys crystals have been grown by the Czochralski method. By means of IR spectroscopy at 300 and 77K temperatures we have determined the carbon content and its distribution in Si and SiGe alloys crystals. Obtained regularity of carbon distribution in SiGe alloys should be explained by high activity of carbon relatively to Si and slight activity relatively to Ge.

1. INTRODUCTION

The SiGe alloys have found wide application in many modern appliances as high-frequency, analog, digital and other microelectronic and photonic devices. An electrically neutral carbon impurity in SiGe alloys among a great number of alloys of doping impurities attracts attention from the technological point of view and a basic science perspective, in order to provide device properties.

As usual, carbon as unforeseen impurity gets into semiconductors because of technology equipment at their obtaining. Pure and doped SiGe alloys crystals are usually grown by the Czochralski growth method. In the process of these crystals growing carbon together with oxygen goes into them. Heated graphite reacts with quartz and in the result silicon carbide compound and carbon dioxide originates. Gaseous products of the reaction come in contact with the melt surface and contaminate it. At the same time in the mutual reaction refractory products are precipitated on the melt surface, react with it and the melt will be enriched with carbon. At the melt cooling carbide compounds are isolated. It has been assumed that carbon which got into the melt participates in complexes deriving process and interacts with defects.

Carbon goes into silicon and germanium lattice mainly by substituting [1]. But behavior of carbon is different with the constituents of SiGe alloys Si and Ge. Generally, the presence of carbon in bulk crystals changes the structure perfection, homogeneity, lifetime of charge carriers making their parameters unstable. But on the other hand carbon is widely used for preparation of modern semiconductor devices including, heterostructures, super lattices, quantum structures on the base of the material system SiGeC [2-7].

Carbon to the SiGe system allows the formation of the ternary alloy SiGeC where one C atom compensates the strain of about ten Ge atoms [3,4]. This allows the growth of layers with increased thickness and Ge concentration while reducing the number of defects. Furthermore, substitution carbon incorporation into the SiGe lattice allows the independent control of strain and Ge composition giving larger choices for the band gaps and band offsets than the SiGe system.

Carbon is constituent of ternary alloy $Si_{1-x-y}Ge_xC_y$ strained epitaxial layers [6]. These materials are used for infrared photo detectors and electronic devices, such as heterojunction bipolar transistors, field effect transistors, and resonant tunneling diodes. The effects of carbon on the structural and electrical properties of $Si_{1-x-y}Ge_xC_y$ and $Ge_{1-y}C_y$ alloys grown by the molecular beam epitaxy showed that the addition of carbon increased the hole mobility in GeC compared to pure Ge. Current-voltage characteristics of SiGeC/Si and GeC/Si heterojunction diodes showed that with increasing carbon, the reverse leakage current decreased and the forward turn-on voltage [6].

There have been obtained at SiGeC materials investigation that at certain composition of C appears to display ordering of the Ge and Si atoms in the structure. The incorporation of C in the Si planes seems to provide local strain centers which facilitate the formation of this ordered phase. The presence of carbon in $Si_{1-x}Ge_x$ influences the thermal oxides of $Si_{1-x-y}Ge_xC_y$ strained epitaxial layers, influences on the dielectric function of the epilayers, thermal budget due to the strain relief of the layer, in the capacitance response of MOS devices.

From the above mentioned follows that the incorporation of carbon into the SiGe alloys system provides an additional degree of freedom which potentially allows for even more versatile band gap engineering, and there is expected to get unusual physical properties due to the large size difference between C and the host atoms in SiGe alloy. Carbon influence on the properties of semiconductors remains unclear up to now.

So it is obvious that investigation of carbon activity in SiGe alloys is of top interest. There are few data in literature under this problem, especially for SiGe alloys. In these work we try to focus on this question. The bulk crystals of Si and SiGe alloys were used for these studies in order to understand the properties of ternary alloy SiGeC system.

2. Experimental Details

Usually Si-Ge alloys are generally obtained as thin films grown on crystalline substrates heterostructures, superlattices and quantum structures. There observed many barriers to the investigation of intrinsic properties of SiGe alloys. To reveal the intrinsic properties of SiGe alloys it is necessary to study properties of their crystals in bulk form.

The bulk crystals of Si and SiGe alloys in the composition range of 0-25 at% Ge were grown by the Czochralski method from quartz crucible with a large volume (exceeding several times the volume of final ingot, in order to minimize composition gradient of the melt)[9]. The fusion was carried out in the helium atmosphere at a pressure of 50 kPa. High purity crystals of n- Si (n≈1011 cm-3) and p-Ge (p≈1013cm-3) were charged into crucible. The composition of SiGe alloys and the uniformity of experimental samples were determined by X-raying. One of the crystals of Si was specially doped by carbon.

Samples were cut from crystals in the form of parallel-sided plate. The surfaces of samples of both sides were mechanically polished by diamond paste with different sizes of grain to reach high optical quality with deviation from flatness<0.002 cm. High quality preserved after mechanochemical polishing. Investigations of the substitutional carbon impurity vibration modes were carried out by measurements of infrared spectra on the UR-20 spectrometer (absolute method measurements) and on the Fourie Transform Infra Red spectroscopy (differential method measurements) and determination of carbon content in experimental samples was implemented by using optical data and the results of activation analysis [8]. The researches were carried out at the room and liquid nitrogen temperatures in the range of 570-660 cm-1. All samples were p-type with current carriers concentration~1015cm-3.

3. Results and Discussion

Silicon. The results clearly show the presence of local mode of carbon atoms vibration in Si (Figure 1.).We observed the absorption band at ν=607cm-1 wave length, corresponding to optically active carbon atoms in silicon. It is known, that optically active are those carbon atoms which take up position in lattice sites of substitution. So the presence of band at ν=607cm-1 wave length in silicon is explained by localized vibration of carbon atoms in substitution sites. The model carbon-substitution impurity is most likely, than interstitial impurity [8].Carbon, replacing Si atom, forms peculiar stretched molecule of silicon carbide. The enlarged distance between carbon atoms and Si reduces strength constants in comparison with ordinary silicon carbide.

It is obvious that cooling of samples from room to liquid nitrogen temperatures causes the shift of all absorption bands by 0.5 % to the higher frequencies.

In this region of spectrum, in particular, at ν=607cm-1 near the carbon band, the absorption band of silicon lattice is observed. To order to exclude the crystal lattice absorption effect, optical measurements in silicon were carried out by using of the differential method. As optically active carbon concentration in Si is proportional to its absorption coefficient maximum (αmax), carbon content in the samples was calculated by the equation N=c ·αmax, where c=1.1·1017cm-2 is a coefficient of graduation, which is obtained by comparison of optical data and the results of activation analysis [8]. The results obtained by the absolute method measurements were checked on the base of the differential method of researches.

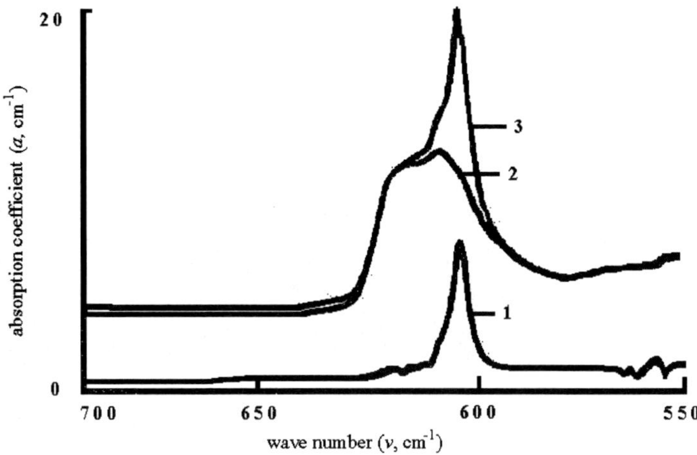

Figure 1. Dependence of absorption coefficient (α) of pure silicon (1) and expressly doped by carbon Si obtained by the absolute (2) and (3) differential methods of measurements vs wave number(ν).

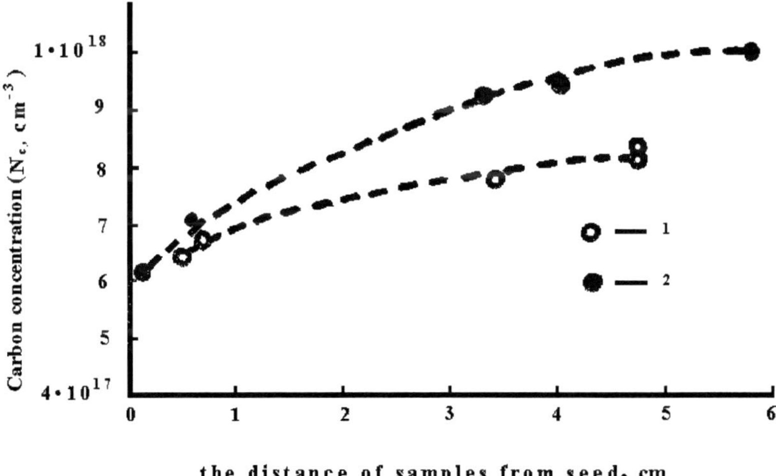

Figure 2. Distribution of carbon concentration (N_c) along the length of Si crystal ingots in undoped(1) and specially doped by carbon(2) Si crystals (starting from the seed end up to the tail end of Si crystals). Dots refer to Si slices cut from crystal ingot.

As an example we cite the distribution of carbon absorption bands obtained by the differential and absolute methods measurements in pure silicon and expressly doped by carbon Si (Figure 1).

Results of carbon distribution investigations in Si are given in Figure 2. Appropriateness of carbon distribution change along the length of undoped and expressly carbon doped crystals is identical. It should be noted that carbon content significantly changes along the crystals length. The value of carbon-doped crystal concentration is more than that of undoped one (Figure 2). Carbon concentration increases at the end of a crystal. It indicates enrichment of the melt with carbon due to segregation during the growth of the crystal, which can not be explained only by the low distribution coefficient of carbon in silicon .It is also connected with collection of carbon in the melt in the result of chemical reactions of silicon with quartz and quartz with graphite. So, we must assume that together with carbon segregation process in silicon the quality of the melt enrichment with carbon will depend on the time of crystal growth process.

Silicon-Germanium Alloys

The experimental optical measurements show that in SiGe alloys there observed the same carbon local vibration modes as in Si in the form of absorption band near ~ 607 cm^{-1}, bonded to Si atoms. In SiGe alloys crystals there was revealed the following law. At formation of SiGe solid solution there have to be observed shift and broadening of absorption band at 607cm^{-1} in the direction of lower frequencies because of presence of heavier atoms of Ge. But because of broadening the shift of the absorption band at increase of Ge content in Si-Ge alloys is not practically discernible. On the whole at decreasing of temperature the maximum of absorption band increases and shifts to higher frequencies.

We have investigated how the concentration of Ge influences on carbon concentration in Si. Analyzing the Si-C vibration bond at 607 cm^{-1} by subtracting out the Si two-phonon peak, we have found that the substitution carbon incorporation in SiGe is a function of Ge content.

Carbon content is maximum in Si and it decreases by increase of Ge concentration(Figure 3). There have not been observed localized modes due to carbon atoms bonded to Ge atoms. Appropriateness of carbon distribution in SiGe alloys can be explained by the great activity of silicon to carbon in relation with Ge. It is stated, that carbon reacts with silicon more actively than with germanium. On the contrary to carbon active interaction with Si, Ge is chemically stable in respect of carbon even at higher temperatures. Germanium does not react with quartz up to 1500^{0}C [9]. Absorption bands in

Ge induced by the presence of carbon have been much less extensively studied than in Si. There have been few reports of localized modes due to carbon in germanium and it seems likely that the carbon solubility is extremely low. Unlike Si and Ge which mix well, the GeC binary system does not occur under equilibrium conditions. Carbon, which has a much smaller lattice constant and atomic radius, is difficult to get incorporated into the crystal lattice substitutionally in Ge. There is an energy preference for substitutional carbon in SiGe alloys to form bonds with Si atoms rather than Ge atoms [11, 12].

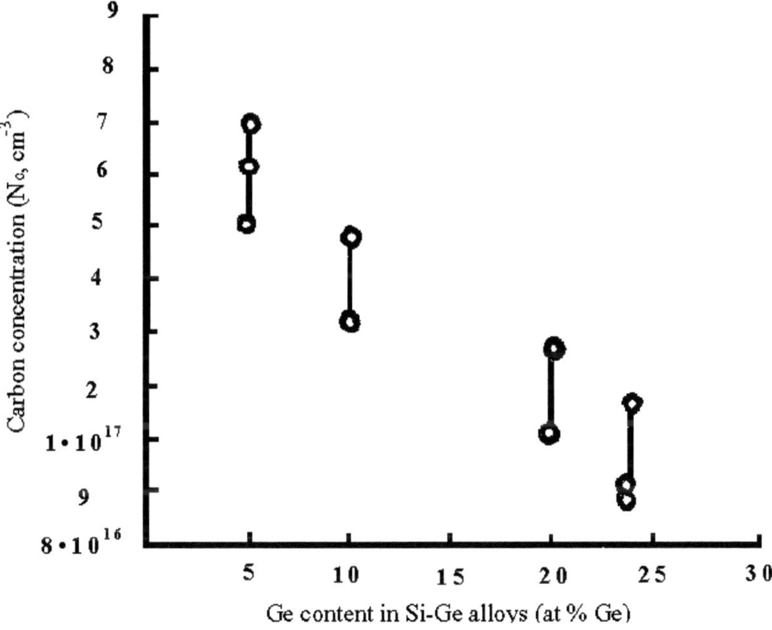

Figure 3. Carbon concentration (N_c) dependence vs. Ge content for Si-Ge alloys.

CONCLUSION

The substitution carbon impurity vibration modes in Si and Si-rich SiGe alloys crystals have been investigated by using Infra Red spectroscopy at 300 and 77K. The carbon content and distribution in these crystals were determined. There was observed the decreasing of carbon concentration with increasing of Ge content in SiGe crystals. Carbon has a preference to substitution in SiGe alloys to form bonds with Si atoms rather than Ge atoms.

REFERENCES

[1] Newman R.C. *Adv. Physics* !969, №T-5,vol.18,sept.

[2] Ishizaka, Akitoshi , Shimada, Toshikazu, United States Patent 4885614.

[3] Amour A. S.et.al. *J. Electronic Materials* 1997, vol.,26,N 12, pp. 1371-1375.

[4] John S, et.al, 38th Electronic Materials Conference,·Santa Barbara, California June 26-28, 1996.

[5] Cuadras A. et.al *Microelectronic Engineering* 2004,vol. 72, Issues 1-4, April, Pages 185-190.

[6] Chen1 F, et al.*Chem. Mater* 2007, May 10.

[7] Todd M,et al. *Chem. Mater* 1996, 8 (10), pp 2491–2498.

[8] Alexandrov G.I *Scie. Proc. GIREDMET* 1971, vol.33,p.170.

[9] Krasiuk B. A., Gribov A. I. *Germanium and Silicon Semicondutors.*M.1961.

[10] Johnson F.A *Proc. Phys. Soc.* 1959,vol.73,pt2,N470, p365.

[11] Hoffman L.,et al. *Phys Rev.* 1999,B60,135733.

[12] Khirunenko L.I. et al, P*hysica* B2007,doi:10.1016/physb.2007.08.146.

In: New Developments in Material Science ISBN 978-1-61668-852-3
Editors: E. Chikoidze and T. Tchelidze

Chapter 11

PLASMA ANODIZING GAAS, WITH APPLICATION OF ULTRA-VIOLET IRRADIATION

A. Bibilashvili*, A. Gerasimov and Z. Kushitashvili
Tbilisi State University, 13, I.Chavchavadze ave.,
Tbilisi, 0179, Georgia

ABSTRACT

A process and a mechanism of stimulation of a low-temperature plasma anodization using ultraviolet radiation in the case of the formation of oxide films of GaAs is proposed. Insulation of the active elements of integrated circuits based on intrinsic GaAs oxide was performed, which provides a significant decrease of leak currents, increases the thermal stability and breakdown field. The stimulating effect of ultraviolet radiation on the process of plasma anodization is attributed by the appearance of additional concentration of antibonding quasiparticles (electrons and holes), which weakens the chemical bonds in the GaAs.

Keywords: Plasma, anodization, stimulation, ultraviolet, intrinsic oxide.

---

* E-mail: amiranbib@yahoo.com

1. INTRODUCTION

Progress in microelectronics and nanoelectronics stimulated the design of fundamentally new methods for producing oxide films, as well as the further development of conventional methods. High-temperature treatment leads to the spread of diffusion regions, the diffusion of unwanted impurities, the generation of defects (dislocations, voids, cracks), the violation of the ratio between the components in binary semiconductor compounds, etc. All these factors have a detrimental effect on the parameters of integrated circuits (IC) and the yield of devices. Taking into account the factors above, there is a reason to believe that low- temperature methods are very promising. Despite this circumstance, the low-temperature process of plasma anodization of metals and semiconductors has not found wide use for the formation of insuiator layers owing to its low efficiency and the low growth rate of oxides [1]. A factor of key importance in GaAs-based technology is the high quality of insulation between active elements of integrated IC.

In this paper we report the results of studying the plasma anodization of GaAs, point to the application of the obtained oxide films for insulating the active elements of IC and suggest mechanisms of low- temperature plasma anodization, which increases the oxidation efficiency and the growth rate of the oxide. Stimulation is attained by ultraviolet (UV) radiation in the course of the plasma anodization [2].

2. EXPERIMENTAL

Wafers with epitaxial *n*-GaAs (100) layers with a electron concentration of $2\times10^{17}sm^{-3}$ were used as substrates. Prior to the technological processes, the surface of the substrate was treated chemically using conventional methods.

Figure 1 represents a plasma anodization system using an UV source. The plasma anodization process is carried out as follows: after achieving the pressure of $1.3\cdot10^{-4}$Pa under the cap (1) the sample (2) is heated by the heater (3) at the temperature range of (373 – 423)K, then argon and oxygen gases with the ratio of 1:2 are fed up to the working pressure P= 6.6Pa; the cathode (4) is heated; the voltage is applied between the anode (5) and cathode; oxygen-containing plasma is formed in the plasma holder (6). Then the surface of the sample is UV illuminated by the UV sourse (7). Finaly, positive bias is applied to the sample with respect to plasma.

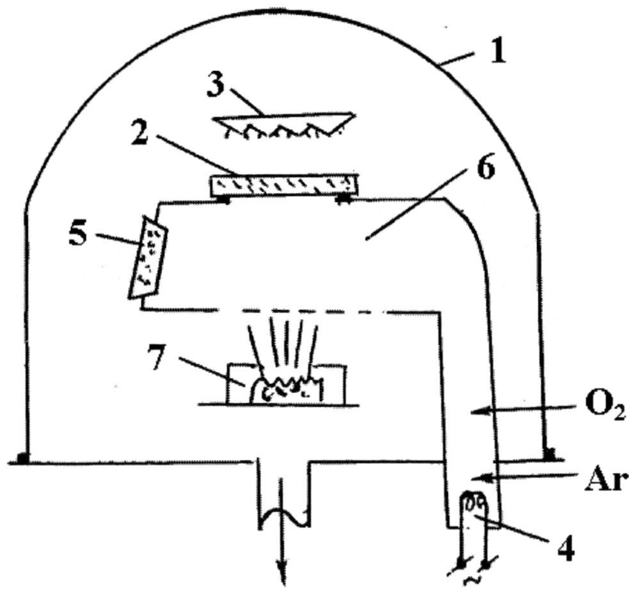

Figure 1. Schematic reprsentation of a plasma anodization system. 1 – cap; 2 – sample; 3 – heater; 4 – cathode; 5 – anode; 6 – plasma holder and 7 – UV radiation source.

When the volitage that is positive relative to the plasma is applied to the sample, the mutual migration of cations of the anodized material and oxygen anions from the plasma occurs, and the oxide of the anodized material is formed as a result of a chemical reaction [1]. The oxide film thickness measured by optical methods was about 1 μm. The process of plasma anodizing assisted with UV irradiation was performed in galvanostatic regime at a substrate temperature of 450 K. The highest energy of the UV photons was 5 eV, so the surface of the anodized materials was not damaged.

Metal-insulator-GaAs structures were formed by depositing aluminum contacts through a mask on the obtained oxides. High-frequency capacitance-voltage (C-V) characteristic measurements and Auger spectroscopy were performed.

3. Results and Discussion

The effect of UV radiation on growth kinetics of the GaAs intrinsig oxide (IO) is illustrated in figure 2. It is noteworthy that the use of UV radiation in the course of plasma anodization not only promotes oxidation but also increases the uniformity of the thickness of GaAs oxide films and reduces the built-in charge in it.

In figure 3 distribution of elements in GaAs IO along the depth, obtained by Auger-spectroscopy is shown. From the data of Auger-spectroscopy and X-ray the analysis follows, that received oxide films are amorphous and in GaAs IO xGa2O3+yAs2O3 x=0.64 and y=0.36. It is necessary to note, oxide films are thermodynamically stable and values of x and y are in a good conformity with the other data [3].

Investigation of high-frequency C-V characteristics (figure 4) showed, that at UV stimulated plasma anodizing, in oxide the fixed charge has decreased from $1.3x10^{11}$ sm^{-2} up to $0.9.10^{11}$ sm^{-2}.

The relation $[Ga_2O_3]/[As_2O_3]=0.64/0.36$ in GaAs IO explained as follows: as Ga atoms are more mobile, than As [4], the probability of their presence in interstitial positions should be more, than the one for As. Accordingly, during GaAs oxidation the probability of interaction of oxygen with the interstitial Ga is higher, than with interstitial As.

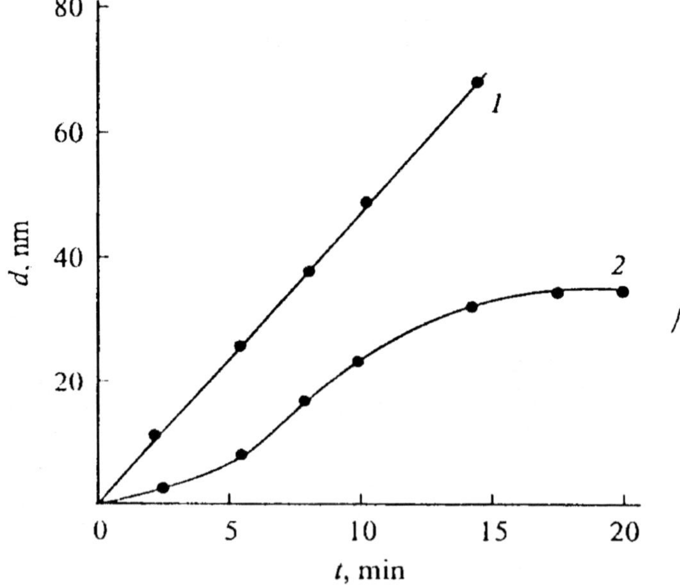

Figure 2. Kinetics of oxidation of GaAs (1) in the presence of UV radiation in the course of plasma anodization and (2) in the absence of this radiation. The forming-current density was I_f= 0,8 mA/cm^2, and the substrate temprtature was T_s=400K.

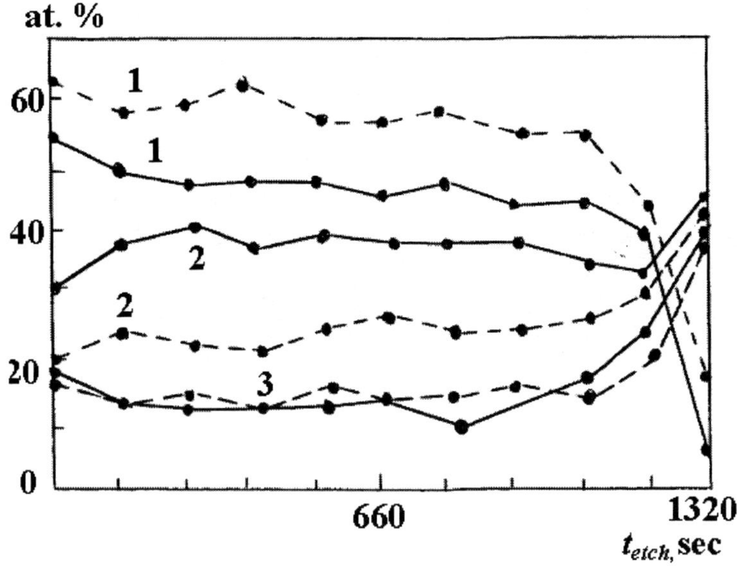

Figure 3. Distribution profile (Auger-spectroscopy) of elements (1-O; 2-Ga; 3-As) of plasma anodizing GaAs (--- without stimulation of process; —— with UV stimulation).

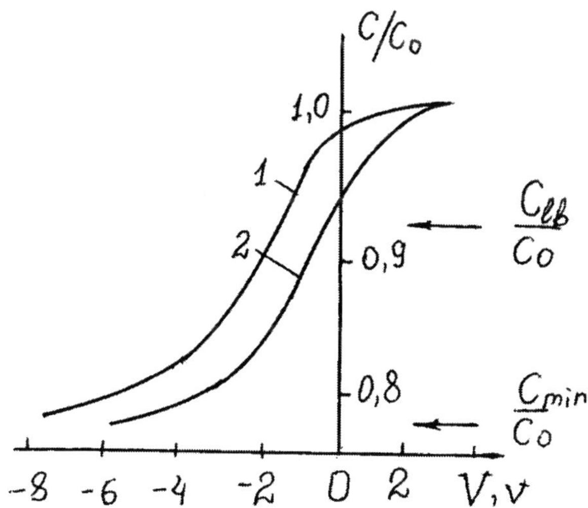

Figure 4. V-C characteristics of GaAs IO: 1 - without stimulation of process; 2 - with UV stimulation.

It should be also noted, that the insulation of active elements using intrinsic GaAs IO is characterized by high thermal stability, ensures high breakdown field (twice as much than that reported in [5]), and also is adapted to planar IC technology, which guarantees the increase of the degree of integration.

Advantages of insulation using intrinsic GaAs IO are explained both by high resistivity of thin oxidefilms, which increases the insulating effect, and by the fact that fixed negative space charge formed in this oxide leads to the formation of an electron- depleted region along the entire boundary with GaAs substrate. This depletion additionally increases the resistance to electron current over the entire boundary.

We attribute the increase in the growth rate to the irradiation of the anodized surface by UV photons, which create antibonding quasiparticles mainly in the bulk GaAs IO, and at the semiconductor-oxide interface too. These quasiparticles weaken the chemical bonds and facilitate the motion of ions. Irradiation by UV photons makes it possible to perform technological processes related to the breaking of chemical bonds at relatively low temprtatures.

The improvement in the uniformity of thickness of the GaAs IO is related to the fact that the surface concentration of nonequilibrium charge carriers generated by UV radiation in GaAs exceeds the doping impurity concentration, which eliminates the effect of nonuniformity of the resistivity over the sample surface.

CONCLUSIONS

1. The increase the concentracion of antibonding quasiparticles in the material, which is anodized as a result of irradiation with UV photons increases the mobility of drifting atoms, which leads to the increase of anodization rate;
2. In order to improve the uniformity of thickness of GaAs IO, it is necessary to irradiate the sample with UV photons in the course of plasma anodization;
3. This method of insulation offers a promising direction in the develompent of GaAs-based IC.

REFERENCES

[1] V. P. Parkhutik and V. A. Labunov, *Plasma Anodization: Physics, Technoiogy and Application in Microelectronics* (Nauka i Tekhnika, Minsk, 1990), p. 276 [in Russian];

[2] S. V. Baras,ev, A. P. Bibilashvili and A. B. Gerasimov, Inventor,s Certificate No. 1172 412 (1985).

[3] C. D. Thurmond, G. P. Schwartz, C. W. Kammlot, GaAs oxidation and the Ga-As-O equilibrium phase diagram, *J. Electrochem. Soc.*, 1981, v.128, # 6, p. 1386;

[4] Y. C. Wang, M. Hong, J. M. Kuo et.al, Demonstration of submikron depletionmode GaAs MOSFET with negligible drain current drift and hysteresis, *IEEE Electr. Dev. Lett.,* 1999, v.20, # 9, p. 457;

[5] E. Y. Chang, C. S. Fuh, C. Cc. Meng, K. B. Wang, S. H. Chen A planar gate double implanted GaAs power MESFET for low voltage digital wireless communication application, *IEEE Trns. Elektr. Dev.*, 2000, v. 37, # 6, p. 1134.

In: New Developments in Material Science ISBN 978-1-61668-852-3
Editors: E. Chikoidze and T. Tchelidze

Chapter 12

FERROMAGNETIC METAL/GAAS HETEROSTRUCTURE

O. Kvitsiani, D. Laperashvili, T. Laperashvili and I. Imerlishvili
Institute of Cybernetics
Sandro Euli str. 5, Tbilisi 0186, Georgia

ABSTRACT

Thin ferromagnetic films (Fe, Ni) on GaAs substrates have involved as a model system for the integration of magnetic materials with semiconductors. For practical applications, it is highly desirable that the injection of spin currents should be electrical and the injecting from a classical ferromagnetic metal in a metal/semiconductor heterostructure is most direct way for spin injection. In majority of investigations metallic thin films has been deposited at elevated temperatures, however, due to the diffusion of Ga and As into the film even at room temperature no sharp interface, satisfying requirements of spintronic devices, is formed. In order to reduce intermixing effects, Fe was recently deposited also at room temperature, and good epitaxial growth was reported without formation of a dead magnetic layer. Here we report the physical properties of structure obtained via electrochemical deposition metallic (Fe, Ni, Pd) thin films on GaAs substrate. Electrochemical deposition is a low-energy process and offers inexpensive alternative for potentially fabricating abrupt Ferromagnetic Metal/GaAs interfaces. Thin ferromagnetic films on GaAs substrates were obtained by electrochemically deposition of ferromagnetic metal on the previously electrochemically etched surface of semiconductor. The aqueous solution of chloride was used for deposition of metals. There were studied current-voltage and capacitance-voltage characteristics for determination interfaces quality, and the results of investigation physical properties of interface are presented in this paper.

1. INTRODUCTION

Deposition of metallic thin films on semiconductor surface is an important subject in electronic technology because every semiconductor device needs the contact to communicate with outside circuits. In our earlier works [1, 2] were investigated Schottky barrier formation mechanism on contacts Metal/GaP and experimentally was obtained, that behavior of Schottky barrier height from metal affinity shows significant deviation from results purposed by Schottky model as well as a Bardeen's model [3]. Nonlinear behavior was explained by interaction of the contacting materials at the interface even at room temperature, especially, for III group metals [4]. Sometimes is question whether is it in reality Schottky contact or there is heterojunction on the boundary metal and semiconductor.

In this work we extend studying Schottky barrier nature in III-V semiconductors and here is described the properties of surface barrier structure based on n-GaAs. Last time interface between ferromagnetic metal and semiconductor is field of research in considerable expansion, because spin transport in metals and semiconductors has great research interest not only for being basic solid state physics issues, but also for the already demonstrated potential to be widely applicable in electronic technology. For deposition of metals on GaAs is used electrochemical method [5]. Electrochemical deposition offers some advantages with respect to ultra-high vacuum (UHV) techniques for the growth of these structures, such as the low temperature processing, which can limit the inter-diffusion between the film and substrate, thus creating well defined interfaces. Electro-deposition is a low-energy process and offers inexpensive alternative for potentially fabricating abrupt Ferromagnetic Metal/GaAs interfaces.

The electrochemical growth of Fe and Ni on (100) oriented n-GaAs was studied in [6-9]. However, due to the diffusion of Ga and As into the film even at room temperature no sharp interface is formed with detrimental consequences for the magnetic moment of the first few nanometers.

The Fe/GaAs contact has attracted intense research interest both because of its magnetic properties and the small lattice mismatch of metal and semiconductor. Formation of Schottky barriers, as well as metallic interconnects, are strongly influenced by the structure and disorder of the interface.

2. EXSPERIMENTAL DETAILS

Deposition of different metals on semiconductor A3B5 was done in a united technological process by electrochemically deposition of metals from aqueous salt solution on the preliminarily electrochemically etched surface. Etching of semiconductor surface occurs just before deposition of metals from the solution, which besides of metallic ions contained electrochemically etching material too.

Samples used for the fabrication of M-S diodes are grown by Chochralski method undoped n-type GaAs (100) oriented wafers. The thickness and carrier concentration was 200-250 μ and 2*1016 – 4*1017 atom/cm3 respectively. At the first ohmic contact to the one side of wafer was formed by alloying of indium at the temperature 500 0C for 5 minutes in hydrogen. Then the sample with ohmic contact and wire for current flow was covered with chemical stable polystyrene solution except the area where the metal to be deposited. The wafers were then etched chemically, rinsed in distilled water and were transferred immediately into electrolyte.

For deposition of contacts were used the chlorides of metals. As it is known for electrochemically etching of A3B5 semiconductors can be used aqueous solution of NaOCl. So, in our experiments as an electrolyte have been used the aqueous solution of chlorides which contained also NaOH and HCl in an amount that the acidity of solution pH =1.5; This electrochemical method is simple and gives possibility of cleaning semiconductors surface before deposition of metal. It is sufficient for fabrication M-S contacts without intermediate oxygen films what is necessary for many applications.

Deposition of ferromagnetic metals Fe, Ni and Pd also were made by the electrochemical method. Electrolyte was poured into quartz glass. The semiconductor wafer was used as a one electrode and as another electrode was used nickel or platinum for deposition of Ni or Fe and Pd. The distance between the electrodes was about 1 cm.

At the first, semiconductor wafer was used as the anode and cleaning of semiconductors surface was done. Then the potential was immediately changed in opposite direction and deposition of metal on freshly cleaned surface was done in the same solution by a united technological process.

Platting up to the thickness of the order 200-500 A0 was obtained within 5-15 minutes. The containment of the aqueous solutions used in our experiment are given in table 1.

Table 1. Metal deposition on semiconductors by electrochemical method

Metal	Composition of electrolyte	Amount g/l	Anode	PH
Fe	FeC13	60	Pt	1.5
Ni	NiCL2.6H2O	30	Ni	1.5
Pd	PdC12	32 - 35	Pt	1.5

After the process of metallization the samples were washed in distilled water. The polystyrene film was removed mechanically and boiled in acetone. Then samples were cut into pieces of area 1-3 mm2. For measurement of electric (Current-Voltage – I(V) and Capacitance-Voltage – C(V)) and photoelectric characteristics presser contacts were used. For investigation of obtained structures Bette's thermo-ionic emission theory [4] was used.

I(V) characteristics were measured in dark at room temperature and C(V) characteristics were measured at a frequency of 1-10 MHz as a function of applied reverse voltage using automatic capacitance bridge model L/7 and L2/28. The spectral response of the diodes was measured at zero bias, using equipment which was constituted with monochromator DMR-4, light source, synchronic detector, frequent-amplifier and recorder.

3. RESULTS

The results of investigation of the electrical and photoelectrical characteristics of Schottky diodes obtained by electrochemical deposition of metals on n-GaAs surface grown by Chokhralsky method are presented. The electrical and photoelectrical characteristics were analyzed by the Thermoelectric emission theory. The barrier heights were determined by the I(V) and C(V) techniques.

The current-voltage characteristics were measured in the stationary regime. According to the theory of thermo-ionic emission model the current flow in Schottky diode gives by following expression:

$$I = I_S\left[\exp\frac{qV}{nkT} - 1\right] \tag{1}$$

where V is the applied forward bias; I – the measured current at a given value of V, q – the electron charge; k – Boltzmann's constant; T – the temperature and IS – is the saturation current density:

$$I_S = A^* S T^2 \exp\frac{\Phi_B^{IV}}{kT}. \tag{2}$$

The barrier height was obtained from the expression (2), by the formula:

$$\Phi_B^{IV} = \frac{kT}{q}\ln\frac{SA^*T^2}{I_S} \tag{3}$$

where S is the area of contact, A* effective Richardson's constant. IS – is the saturation current, which is obtained by the interpolation of relation (V, lnI), when V=0.

The value for the ideality factor (n) of diodes was calculated by the equation:

$$n = \frac{q}{kT}\frac{\Delta V}{\Delta \ln I} \tag{4}$$

In Figure 1 is shown a typical Current-Voltage characteristic of Fe/GaAs Schottky diode for forward bias. As can be seen (lnI, V) dependence for a little forward bias is strait line with ideality factor n=1.02 and barrier height can be calculate by expression (3) and there was obtained:

$$\Phi_B^{IV} = 0.80ev \tag{5}$$

For comparison the barrier heights obtained by different method was studied capacitance-voltage characteristic too. Junction capacitance was measured at a frequency of 1-10 MHz as a function of applied reverse voltage using automatic capacitance bridge models L/7 and L2/28. Graph in (V, 1/C2) dependence was linear and C(V) data were analyzed according to the (C, V) relation for an ideal contact. When the space charge are uniformly distributed in a surface depletion region, the diffusion voltage is determined by a Capacitance–Voltage characteristics.

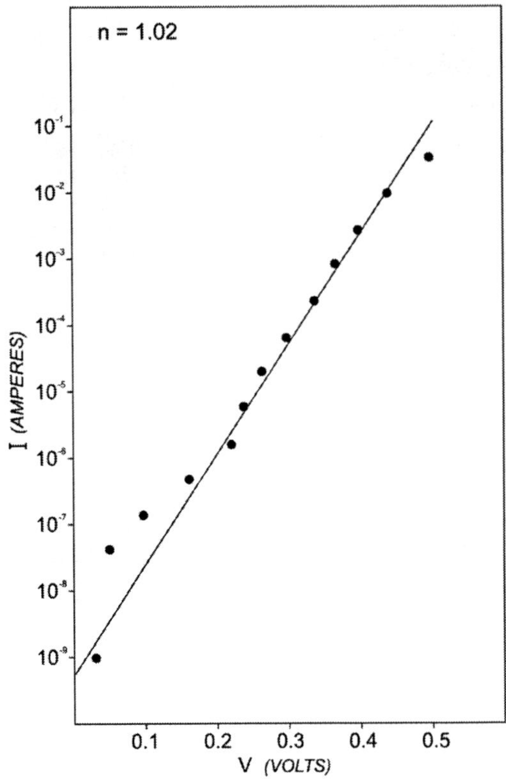

Figure 1. Current–voltage characteristics of Fe/GaAs diode, for forward bias voltage.

Extrapolating the data of C-2 into the voltage axis V0:

$$\frac{1}{(C^2)} = 2(\varepsilon q N_D)^{-1}(V_d - V_0 - \frac{kT}{q}) F^2 cm^4 \qquad (6)$$

where the intercept Vo on the voltage axis is related to the diffusion potential Vd by

$$V_0 = \left(V_d - {}^{kT}\!/_{q}\right)$$

and ND is the bulk donor concentration. The C-V measured barrier height is

$$\varphi_B^{CV} = V_0 + E_F + \frac{kT}{q} \tag{7}$$

where EF is the Fermi Energy in the bulk with respect to the conduction band. For the material used:

$$\varphi_B^{CV} = V_0 + 0.08ev.$$

The measurement accuracy of φ_B^{IV} and φ_B^{CV} for the tick contacts is estimated to be <0.01eV. In Figure 2 is given Capacitance-Voltage characteristic of obtained structure Fe/GaAs, and can be seen that the values of Schottky barrier height obtained from Current-Voltage φ_B^{IV} and Capacitance-Voltage φ_B^{CV} characteristics are in a good agreement, suggested, that prepared contact Fe/GaAs is abrupt.

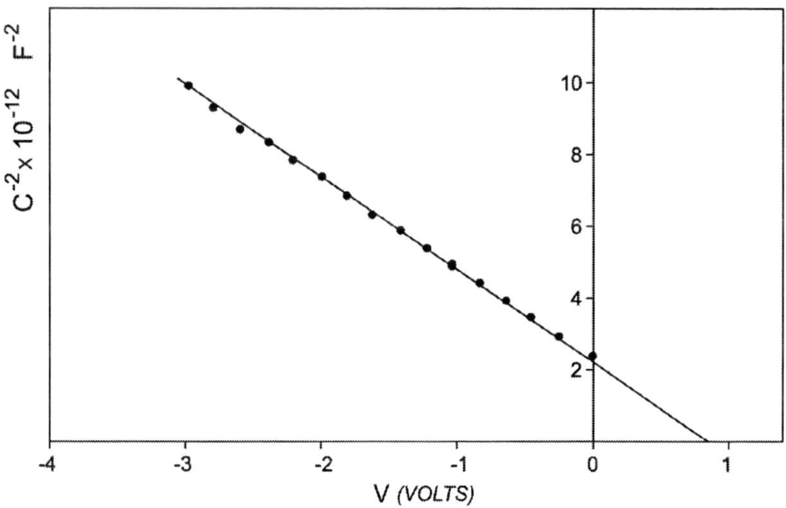

Figure 2. Capacitance-voltage characteristic of Fe/GaAs structure.

Schottky barrier width W0 was measured by equation:

$$W_0 = \frac{\varepsilon_S \varepsilon_0 S}{C_0}, \tag{8}$$

where ε_S - is dielectric permittivity of semiconductor (12.9 for GaAs); ε_0 - dielectric permittivity of vacuum (8.85*10-12); S – contact area; C0 – initial capacitance. Obtained width was: W0 = 1.4*10-6 cm.

The results of investigation of electrical and photoelectrical characteristics of Schottky diodes In/GaAs, Ga/GaAs, Al/GaAs obtained by electrochemical deposition and Schottky diodes Au/GaAs and Ni/GaAs prepared by chemical deposition of metals on n-GaAs surface grown by Chochralsky method, with current concentration 2*1017cm-3 were presented in [2]. The electric and photoelectric characteristics were analyzed by the Thermoelectric Emission Theory. Barrier heights were determined by the Current-Voltage and Capacitance-Voltage techniques. The both measurement technique gave the close barrier height.

Photo excitation over the barrier (ϕ<hv<Eg) was measured and Fowler plots of photo current spectrum show linear regions extrapolation of which gives the barrier height ϕB (If - hv). By the same way were constructed the characteristics for another samples (Ni, Pd). The characteristics were constructed at the coordinates (V, lnI), (V, I/C2) and (hv, IF1/2) and calculated values of barrier height are given in Table 2:

Table 2. The barrier height of diodes metal/GaAs

Structure	Ideality factor, n	Barrier height		
		J/V	C/V	PR
Fe/GaAs	1.02 –1.05	0.80	0.81	
Ni/ GaAs	1.03 –1.05	0.9	0.9	0.9
Pd/GaAs	1.05 - 1.07	0.9	0.9	0.9
Au/GaAs	1.05 - 1.07	0.90	0.92	0.93
Ga/GaAs	1.05 –1.07	0.90	0.90	0.90
Al/GaAs	1.07- 1.09	0.87	0.80	-
In/GaAs	1.05 –1.07	0.70	0.72	0.7

CONCLUSION

Electrochemically deposited Ferromagnetic Metal/GaAs (100) contact shows abrupt Schottky barrier with ideality factor n=1.02 – 1.07. From investigation of this Fe-n-GaAs Schottky Diode using Bette's Thermo-Ionic Emission Theory was found, that the barrier height of the structures obtained by electrochemical deposition of Fe on freshly electrochemically etched surfaces of Chokhlarsky method grown (100) oriented GaAs ws 0.80 eV. The values of barrier height, obtained by I(V), C(V) and PR behavior, of near-ideal diodes commonly are in a good agreement. But calculated Schottky barrier heights are greater than is known from literature data [9-11].

ACKNOWLEDGMENTS

The designated project has been fulfilled by financial support of the Georgia National Science Foundation (Grant #GNSF/ST08/4-426). Any idea in this publication is possessed by the author and may not represent the opinion of the Georgia National Science Foundation itself.

REFERENCES

[1] Laperashvili, T.; Imerlishvili, I., "Electrochemically deposited Schottky barrier height", *Proceedings of Institute of Cybernetics*, vol. 1, N 1-2, p. 208-211 (2000).

[2] Laperashvili, T.; Nakashidze, G., "Behavior of Schottky barrier height from metal's electronegativity in structure based on GaP", *The Journal of Technical Physics*, vol. 55, N 4, p. 733-734 (1985).

[3] Laperashvili, T.; Imerlishvili, I.; Khachidze, M.; Laperashvili, D., "Photoelectric charachteristics of contacts in A3B5 semiconductors", *Proceedings of SPIE Nano- and Micro- Technology*, (2003).

[4] Rhoderick, E.H., "*Metal-Semiconductor Contacts*", Oxford University, New York (1978)

[5] Laperashvili, T. "*Method of Fabrication of Schottky barrier*", pat. rep.Georgia #46 , priority 27.05 (1992).

[6] Bao, Zhi Liang; Kavanagh, Karen L. "Epitaxial Fe/GaAs via electrochemistry", *Journal of Applied Physics* 98, 066103-1-- 066103-3 (2005).

[7] Erwin, Steven C.; Lee, Sung-Hoon; Scheffler, Matthias. "First-principles study of nucleation, growth, and interface structure of Fe/GaAs". *Physical Review* B, V.65, 205422-1--205422-10 (2002).

[8] Wunnicke, O.; Mavropoulos, P. and Dederichs, P. H., "Spin Injection: Interface Resistance in Fe/Semiconductor Junctions Calculated from First Principles", *Journal of Superconductivity: Incorporating Novel Magnetism*, Vol. 16, No. 1, February 2003 (2003).

[9] Honda, S. Itoh, H. Inoue, J. Kurebayashi, H. Trypiniotis, T. Barnes, C. H. W. Hirohata, A. and Bland, J. A. C., "Spin polarization control through resonant states in an Fe/GaAs Schottky barrier", *PHYSICAL REVIEW* B 78, 245316 (2008).

[10] McLean, A. B.; Williams, R. H.. "Schottky contacts to cleaved GaAs (110) surfaces. I. Electrical properties and microscopic theories", *Journal of Physics C: Solid State Physics*, Volume 21, Issue 4, pp. 783-806 (1988).

[11] McLean, A. B.; Williams, R. H.. "Schottky contacts to cleaved GaAs (110) surfaces. II. Thermodynamic aspects", *Journal of Physics C: Solid State Physics*, Volume 21, Issue 4, pp. 807-818 (1988).

In: New Developments in Material Science ISBN 978-1-61668-852-3
Editors: E. Chikoidze and T. Tchelidze

Chapter 13

INVESTIGATION OF TECHNOLOGICAL PROCESS OF MAGNESIUM SILICIDE PRODUCING BY MAGNESIUM THERMAL RESTORATION OF LOW-GRADE QUARTZITE

Iuza Pulariani, George Darsavelidze, Merab Kereselidze, Roland Razmadze, and Elza Khutsishvili*

Ferdinand Tavadze Institute of Metallurgy and Materials Science, Tbilisi, Georgia

ABSTRACT

The technology of receiving of technical Si or/and magnesium silicide has been described. Presented technology is based on the magnesium thermal restoration method. In the technology offered by us there has been applied low-quality, cheap, local quartzite in which content of $Si0_2$ does not exceed 95% and therefore this quartzite can not be applied without its ore-dressing. The process is strongly exothermal and in the case of selection of suitable conditions the process goes on at the expense of own heat. According to the quantity of added magnesium it is possible to obtain Si, magnesium silicide or both of them. The expenditure of energy is reduced marginally.

---

* E-mail: elzakhutsishvili@yahoo.com

We studied the influence ratio of quartzite/ Mg, fractional composition of quartzite, duration of the process and initial temperature at the restoration process. Values of the noted factors, which provide achievement of maximum degree of technical Si and magnesium silicide purity during the production, have been established.

1. Introduction

Solution of energetic problems, connected with an exhaustion of organic fuel stocks becomes more and more urgent. According to the fulfilled researches by 2020 the organic fuel will satisfy of the world economics demands only partly. Pending of energetic crisis sharpening researchers' attention is directed to a search of alternative sources. Application of such renovated sources of energy as solar energy is perspective. Work on this direction has been carried out since the first half of the XX century. Photoelectric converters (P.E.C.) manufactured on the Si basis are considered as perspective. In many different countries of the world such assembled P.E.C.s has already produced total electric power of 2500 megawatt in 2006. It is expected a quick growth of photoelectric stations world production in the near future. According to specialist estimation, annual global need of solar quality Si after 2010 can make 16000- 32000 tons. However at that the delivery is supposed in amount of 8000 tons. Therefore, search of new sources of "solar" quality silicon is necessary. Moreover Si is widely used in electronics and microelectronics. Here used Si is purer than applied in P.E.C.s so–called "solar" Si.

High price of "solar" quality Si prevents the development of "solar" power engineering up-tempo. Aspects of the development of solar power are narrowly connected to problems of development of polycrystalline silicon producing new technologies with low expenses. In overwhelming majority of cases technical (metallurgical) silicon is initial raw material for reception of semiconductor polycrystalline silicon. Today it is melted in so-called ore thermal electro arc furnace. The process is characterized by a number of essential lacks, in particular by: the high specific inputs of the electric power, high requirements to quality of used raw material and materials. This stipulates the high price of as a result obtained Si. Whereas *50%* of the P.E.C cost *amounts* the price of *material* necessary for P.E.C.s manufactures. Therefore, the efforts of researchers are directed to the cost price reduction of semiconductor-purity Si and to the efficiency increase of the P.E.C.s.

There exists the alternative way of technical (metallurgical) silicon manufacture. For that silicon dioxide is restored by magnesium. By interaction of Si, obtained as a result of restoration, with magnesium there appears magnesium silicide. It represents initial raw material for silane manufacture. By thermal decomposition of it there is obtained semiconductor purity Si. The process of magnesium thermal restoration of quartzite is strongly exothermal and its realization demands of less power inputs than processes of Si obtaining by other methods. Indicated technology has not found spreading because of high cost of restorer (magnesium).

Presented technology foresees magnesium *reception by off-centre method, which is based on the restoration of* magnesium chloride by steel filings. The restoration process is characterized by low *power inputs and accordingly decreasės the price of* obtained magnesium, which is produced to our time by electrolysis of chloride melt.

The restoration process is expressed by equation:

$$3MgCl_2 + 2Fe \rightleftharpoons 3Mg + 2FeCl_3 \qquad (1)$$

The process is reversible and thermodynamically flows at 700^0C *from the right to the left. But because it runs from the left to the right, there originate gas products and arises* magnesium. Indicated circumstance makes perspective technical Si and/or magnesium silicide receiving process by magnesium thermal restoration of quartzite. The restoration process is expressed by equations:

$$SiO_2+2Mg=2MgO+Si \qquad (2)$$

$$SiO_2+4Mg=Mg_2Si+2MgO \qquad (3)$$

This time it is used a low quality, cheap quartzite, in which the content of *SiO2* $_{does}$ *not* $_{exceed}$ *95% and which does not find corresponding application without ore-dressing. At that it is possible to obtain Si,* magnesium silicide *or both together depending on* magnesium quantity. Power inputs of process are minimized.

2. The Details of Experiments

Restoration process is carried out in corundum beaker placed into mine furnace in *argon atmosphere. The temperature was measured in the furnace and reaction area as well. After reaching the preliminarily appointed initial temperature the* beaker with reaction mixture was placed into the furnace. Since the moment of exothermal effect appearance in the reaction mixture we started counting out the duration of the process. The degree of quartzite restoration by magnesium was established by means of chemical and X-ray phase analyses.

3. Results and Discussion

It was studied the influence of the following factors on the process: 1-initial temperature, 2-ratio of quartzite/ *Mg, 3-duration of the process and 4-fractional composition of quartzite.*

The influence of initial temperature on the restoration process has been investigated at 600,650,700,750 and 800^0C by constant values of other factors magnitudes. Results of investigations have shown that restoration process *starts at 700^0C and at 800^0C temperature reaches maximum-90%. At indicated temperature there has been studied influence of other factors on the restoration process. It was taken various quantities of restorer for studying influence of reagent substances ratio, where stoichiometric quantity was minimum according to the reaction equation.*

It was established, that it is possible to obtain Si during the restoration process, when a restorer is taken by excess of 120% with regard to stoichiometric.

The more magnesium is added, the more *quantity of* magnesium *silicide (Mg_2Si) is created in the system together* with *the increase of restorer. When amount of* magnesium reaches 130 % of *stoichiometric quantity, value necessary to create* magnesium *silicide, the metallic phase of system is completely presented by* magnesium *silicide. At that the degree of quartzite transition into* magnesium *silicide is not less than 95%. The study of influence of process duration has shown, that it is enough 20 min. for Si and silicide creation.*

For obtaining of high purity Si from restored mass it is necessary its treatment by hydrochloric and fluoric acids or corresponding halogen hydrogen compounds. At that obtained Si purity is not less 98%.

In the case when magnesium *silicide presents target product its isolation happens by system heating up to silicide melting-point(1085^{0}C) and by its separation from solid phase with special filter. Indicated method can be also used in that case when target is the reception of Si and silicide as well. It is possible to reach this aim by heating of* magnesium *silicide over its thermal decomposition temperature(1100^{0}C).*

Conclusion

According to the performed researches, it should be noted:

- *Obtaining of Si and/or its silicide is possible* from a low quality quartzite with content of Si not less 90% Si *by* magnesium thermal restoration method.
- *Obtaining of* Si, magnesium *silicide or both of them is available according to the added quantity of* magnesium.
- *High pure Si and/or* magnesium *silicide have been received on the base of which we can obtain semiconducive Si by the well known methods.*
- *High purity of obtained products and high economic index are due to the high quality of conversion which had been achieved during the restoration process.*

In: New Developments in Material Science ISBN 978-1-61668-852-3
Editors: E. Chikoidze and T. Tchelidze

Chapter 14

INFLUENCE OF SHAPE OF MAGNETIC PARTICLES ON MAGNETO-OPTICAL PROPERTIES OF THE ULTRAFINE STRUCTURES

Omar Nakashidze * ***and Lali Kalandadze***
Depart. of Physics, Shota Rustaveli State University,
Batumi 6010, Georgia

ABSTRACT

In the present paper, using the discontinuous Fe films as examples we have investigated the influence of the shape of magnetic particles on the magneto-optical and optical properties of the ultrafine structures. Obtained results have confirmed the significant change of the components of the tensor of effective dielectric permittivity and subsequently of the magneto-optical and optical properties which was brought about by the change of the shape of magnetic particles.

These calculations proved that if we take into account the shape of the particles, we will achieve a perfect match of the experimental results and the theoretical calculations.

* E-mail: omari_nakashidze@mail.ru

1. INTRODUCTION

Nowadays, magneto-optical research methods are widely used for examination of ultrafine magnetic structures such as: magnetic fluids, thin discontinuous metal films, heterogenic glasses, ferrite-garnet films, implanted magnetic surfaces and etc. Ultrafine medium is an ensemble of particles having smaller size than 1000 $\overset{0}{A}$. They have unique qualities compared to bulk ferromagnetic. This appears as a result of the fact that the identical properties particles and materials of particles have, are destroyed at a diapason covering size till 1000 $\overset{0}{A}$ [1]. There is no doubt that any experimental or theoretical research done in this direction revives a strong interest as many aspects remain still enigmatic.

2. THEORY

In general, the magneto-optical properties of ultrafine magnetic medium are very different from the properties of the bulk ferromagnetic and depend on the structural parameters: the occupancy of the volume of the ultrafine medium with metal, the size and shape of the particles, the order of the particles, the properties of the medium, surrounding metal particles [2,3].

The tensor of effective dielectric permittivity of magnetized ultrafine ferromagnetic material, by analogy with the tensor for bulk ferromagnetic materials, can be represented as

$$\varepsilon_{\text{eff}} = \begin{bmatrix} \varepsilon_{\text{eff}} & -i\varepsilon'_{\text{eff}} & 0 \\ i\varepsilon'_{\text{eff}} & \varepsilon_{\text{eff}} & 0 \\ 0 & 0 & \varepsilon_{\text{ef}0} \end{bmatrix}, \qquad (1)$$

where $\varepsilon_{eff} = \varepsilon_{1eff} - i\varepsilon_{2eff}$, $\varepsilon_{0eff} = \varepsilon_{01eff} - i\varepsilon_{02eff}$ and $\varepsilon'_{eff} = \varepsilon'_{1eff} - i\varepsilon'_{2eff}$.

In this case the tensor components depend on both the properties of magnetic colloidal particles themselves, and the properties of the medium in which they find themselves. The knowledge of tensor components enables us to calculate any type of magneto-optical effect. The next step is to establish the link between the components of the tensor of effective dielectric permittivity and the components of the tensor of the according bulk materials. The first this task was brought in 1985 [4].

For magnetic fluids with a low concentration of magnetic colloidal particles and with no direct contact between them the tensor components of the effective dielectric permittivity within the framework of theoretical models of an effective medium, extended to include the case of magnetic media, can be written as:

1. The model of averaged characteristics:

$$\varepsilon_{eff} = q\varepsilon_m + (1-q)\varepsilon_0 \ ; \ \varepsilon'_{\text{eff}} = q\varepsilon'_m \tag{2}$$

2. The Maxwell-Garnett model:

$$\varepsilon_{eff} = \frac{2q(\varepsilon_m - \varepsilon_0) + (\varepsilon_m + 2\varepsilon_0)}{(\varepsilon_m + 2\varepsilon_0) - q(\varepsilon_m - \varepsilon_0)} \ ;$$

$$\varepsilon'_{eff} = \frac{9q\varepsilon_0^2\varepsilon'_m}{(\varepsilon_m(1-q) + \varepsilon_0(2+q))^2} \tag{3}$$

3. The Bruggeman model:

$$\varepsilon_{eff} = \varepsilon_o(1 + 3q\frac{\varepsilon_m - \varepsilon_o}{\varepsilon_m + 2\varepsilon_o}) \ ; \ \varepsilon'_{eff} = \frac{9q\varepsilon_o^2\varepsilon'_m}{(\varepsilon_m + 2\varepsilon_O)^2} \ , \tag{4}$$

where $\varepsilon_m = \varepsilon_{1m} - i\varepsilon_{2m}$ and $\varepsilon'_m = \varepsilon'_{1m} - i\varepsilon'_{2m}$ are the diagonal and nondiagonal tensor components of the dielectric permittivity of the material of magnetic colloidal particles, ε_0 is the dielectric permittivity of the fluid phase, and q is the ratio of the volume, occupied by magnetic particles, to the total volume of the magnetic fluid.

After generalization of this theory we arrived at formula how to calculate tensor components for non-spherical ultrafine particles.

$$\varepsilon_{ef} = \varepsilon_0 \left(1 + \frac{q(\varepsilon_m - \varepsilon_0)}{y\varepsilon_m + (1-y)\varepsilon_0}\right) \quad \varepsilon'_{ef} = \frac{\varepsilon_0^2 \varepsilon'_m}{y\varepsilon_m + (1-y)\varepsilon_0{}^2} \qquad (5)$$

where $y = f_2 - qf_1$, whilst f_1 and f_2 are the factors of the shape of the ultrafine particles and their surrounding medium [5].

These formulas (2) can be used to investigate which structural parameter has more influence on the optical and magneto-optical properties of the ultrafine medium.

The equatorial Kerr effect under consideration, which consists in a change in the intensity of linearly polarized light reflected from the sample in the case reversal of magnetization of the sample, can be written as

$$\delta_p = \frac{2\sin 2\varphi}{A^2 + B^2}\left(A\varepsilon'_{1eff} + B\varepsilon'_{2eff}\right) \qquad (6)$$

where

$$A = \varepsilon_{2eff}\left(2\varepsilon_{1eff} f \cos^2\varphi - 1\right); \quad B = \left(\varepsilon_{2eff}^2 - \varepsilon_{1eff}^2 + 1\right)\cos^2\varphi + \varepsilon_{1eff} - 1$$

end φ is the angle of light incidence.

3. EXPERIMENTAL

In this work we investigate the magneto-optical properties of discontinuous iron films, the weight thickness d ($d = m/\rho S$, where m - film mass, ρ-metal density and S-film square) of which falls within the interval 10-300 Å. Discontinuous films were obtained by evaporation in vacuum of 10^{-5} Torr on glass substrates with a rate of 1 to 5 Å/S. For investigating the magneto-optical properties we have chosen the equatorial Kerr effect. The optical constants were determined using the Every method.

4. Results and Discussion

Figure 1 presents the spectral dependences of the equatorial Kerr effect for the discontinuous iron films with different weight thicknesses d . Given there also is a similar frequency dependence for polycrystalline iron.

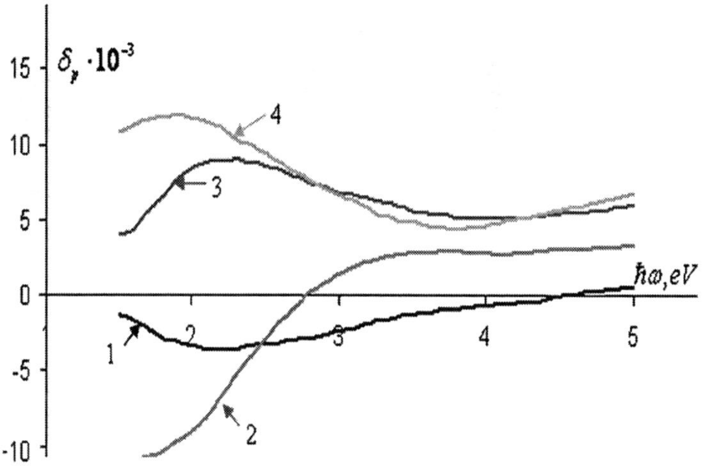

Figure 1. Experimental dependences of the equatorial Kerr effect of $\hbar\omega$ for discontinuous Fe films with weight thicknesses $d = 54$ (1); 75 (2); 112 $\overset{o}{A}$ (3) and polycrystalline iron (4) ($\varphi = 70^0$).

It can be seen from the figure 1 that for the films with a different weight thickness the character of the frequency dependences of the equatorial Kerr effect is significantly different from that of the similar dependences for bulk iron and depends on the film effective thickness.

Obtained experimental results were compared with calculations in the framework of the theoretical Maxwell-Garnett model. Figure 2 gives the dependences of equatorial Kerr effect on $\hbar\omega$, calculated for ultrafine iron medium as given in eqs.(3) and (6) for different q and $\varepsilon_0 = 1$. In this and following calculations the dielectric permittivity tensor components for bulk iron were used.

Analysis of obtained results shows that the theoretical Maxwell-Garnett model qualitatively describes the experimental results. The difference between experimental and theoretical dependences may be explained by influences shape of particles and ε_0. With that end in view we calculated frequency dependences of equatorial Kerr effect by formulas (5) and (6) for different q, f and ε_0=1,5. By way of example figures 3 and 4 present the results of calculations, using equations (5) and (6), of equatorial Kerr effect for different q and f=1/3 and f=1/4.

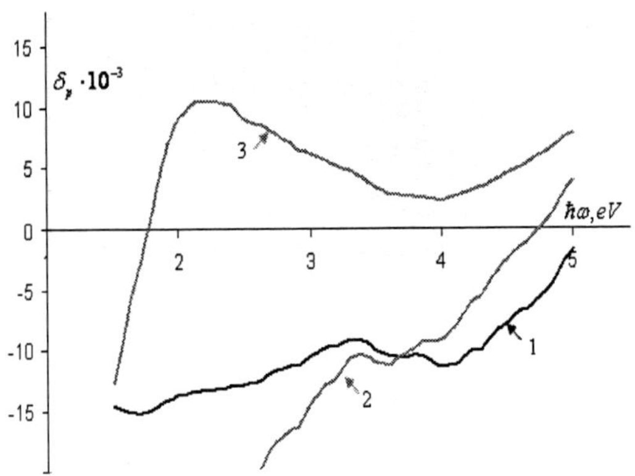

Figure 2 Dependences of the equatorial Kerr effect on $\hbar\omega$ for ultrafine iron, calculated within the framework of the Maxwell-Garnett model with q=0,3(1); 0,5(2) 0,7(3) and $\varepsilon_0 = 1$, $f = 1/3$ (φ=70^0).

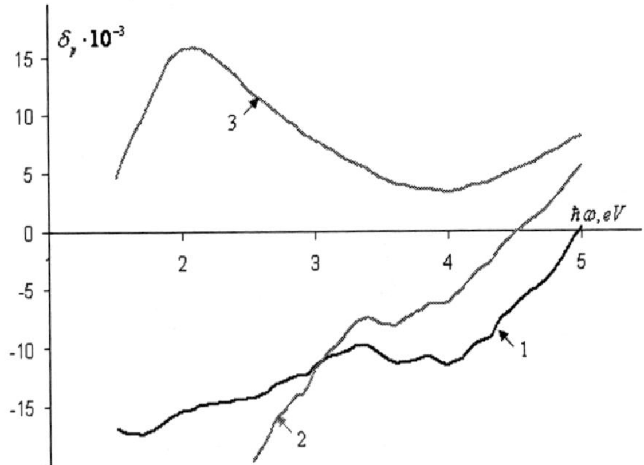

Figure 3. Dependences of the equatorial Kerr effect on $\hbar\omega$ for ultrafine iron, calculated by formulas (5) and (6) with q=0,3(1); 0,5(2) 0,7(3) and $\varepsilon_0 = 1, 5$; $f = 1/3$ (φ=70^0).

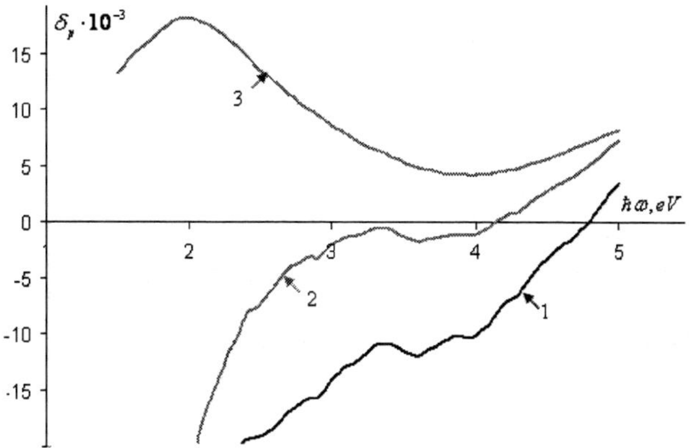

Figure 4. Dependences of the equatorial Kerr effect on $\hbar\omega$ for ultrafine iron, calculated by formulas (5) and (6) with q=0,3(1); 0,5(2) 0,7(3) and $\varepsilon_0 = 1,5$; $f = 1/4$ ($\varphi = 70^0$).

It is evident from the figures 3 and 4 that the shape of magnetic particles influences significantly the magneto-optical properties of discontinuous iron films.

Conclusion

In the present paper, using the discontinuous Fe films as examples we have investigated the influence of the shape of magnetic particles on the magneto-optical and optical properties of the ultrafine metal films. Obtained results have confirmed the significant change of the components of the tensor of effective dielectric permittivity and subsequently of the magneto-optical and optical properties which was brought about by the change of the shape of magnetic particles.

These calculations proved that if we take into account the shape of the particles, we will achieve a good agreement of the experimental results and the theoretical calculations.

REFERENCES

[1] Nikitin L V and Kasatkina O V 1986 Proc. X All-Union school-seminar on *New magnetic materials of microelectronics* (Riga) vol 2 124 (in Russian).

[2] L.V.Nikitin, L. G. Kalandadze, M.Z.Akmedov, S.A.Nepijko, A.P. *Ostranica journal of Magnetism and Magnetic Materials* 148 (1995) 279-280.

[3] L.V.Nikitin and O.M.Nakashidze. *Journal of Magnetism and Magnetic Materials,* 83 (1990) 77-78.

[4] G.S. Krinchik, L.V. Nikitin and O.V. kasatkina, Poverkhnost', *Fizika, Khimiya, Mekhanika* 7 (1985) 140.

[5] O.M.Nakashidze, 1998 Proc. Of Batumi State University, Vol.2, 78.

In: New Developments in Material Science ISBN 978-1-61668-852-3
Editors: E. Chikoidze and T. Tchelidze

Chapter 15

APPLICATION OF SHOCK-WAVE LOADING AND SUBSEQUENT NEUTRON IRRADIATION FOR ARTIFICIAL DIAMOND SYNTHESIS

***I. Chkhartishvili*[1]*, J. Sharashenidze*[1]*, F. Akopov*[2] *and N. Kvataya*[1]**

[1]Tavadze Institute of Metallurgy and Materials Science
[2]E. Andronikashvili Institute of Physics

ABSTRACT

The energetic conditions essential for graphite transformation into the diamond are formed by using the energy of explosion and neutron irradiation.

First the initial metal-carbon system is subjected to loading by weak shock waves (pressure ≤ 1,75 GPa). The graphite is transformed into the activated state with a rhombohedric structure. Further the irradiation with fast neutrons under low-temperature conditions (300K) is performed. The neutron dose are about $3 \cdot 10^{22}$ N/m^2; energy –E ≥0,1mev.

As a result of such complex treatment, the phase transformation of graphite into the diamond is accomplished.

The neutron irradiation is the key stage in the proposed technology. Available reactors can be used for implementation of this stage.

INTRODUCTION

It is known that a carbon is an initial material for natural diamond synthesis, which is a result of a restoration process of carbonate strata by bivalent iron (at high temperatures and pressures) [1].

An original conception was so called Abyssal hypothesis of diamond formation. According to this hypothesis, diamonds are synthesized at a depth of 70-100 km from the Earth surface, temperature- 2000^0C and pressure-50-60kbar. A deficiency of Abyssal hypothesis is that such pressure value is unreachable at a depth of 70-100 km.

The next stage is a formation of explosive hypothesis. According to it, diamonds are formed at a depth of 4-6 km from the Earth surface. An ascending magma can form expanding channels and sites for further diamond synthesis at indicated depth. Diamonds are formed in the moment of explosion of free carbon, which is sufficiently saturated in the site because of the process of carbohydrates dissociation on carbon and hydrogen. Explosive hypothesis has many essential deficiencies. Diamonds occur as kimberlites as well as single crystals. They are more or less uniformly distributed in depth. If diamonds were formed at explosion of concentrated mass of carbon substance, then large diamonds bodies would be found in clusters. Moreover, minerals formed at pressures no more 20-35kbar would be found in kimberlite pipes along with diamonds. It means that kimberlites were not effected a pressure more than 20-35kbar. Therefore, a question is: how is diamond formed along with these minerals required formation pressure of which is more than 50kbar.

An interesting hypothesis is presented in [2]. According to this hypothesis, a total pressure is low in magma, but in some local sites pressure increases up to higher values at the expense of cavitations. Cavitation occurs in flowing liquid with gas dissolved in it, when liquid flows along a channel of variable cross-section at a certain velocity.

Kimberlite magma contains carbonic acid. The smallest bubbles of carbonic acid quickly grow; cavitation bubbles are formed; bubbles begin to collapse in a wider channel area and heavy shrinkage forces appear in a small volume.

A motion rate of molten magma required for cavitation is equal to 300-500 m/sec according to calculation [2]. At this case magma pressure is equal to 20kbar. This pressure is necessary for formation of diamond satellite - Arizona ruby mineral. A pressure in a bubble in the collapse process is equal to 1000kbar (million of atmospheres). This value is averaged. The maximal pressure that can be reached is 5-10 10^6 atm. It was shown [3] that required velocity of molten magma is 300-500 m/sec.

According to cavitation hypothesis, a cavitation pressure can so strongly approach carbon atoms that crystal is initially getting more compact than diamond. Afterwards, during an extension, this substance gets a crystal structure of diamond. In [4] data about existing of insufficiently studied carbon modification- “Carbon -III”, which is more compact than diamond, is presented. Diamond conservation is reached at the expense of adiabatic expansion and cooling of initial most compact substance with following transition into diamond modification. In this case cooling takes not because of heat exchange, but at the expense of instantaneous phase transition.

Mechanisms of diamonds natural and artificial diamond creation are significantly distinguished. High value of parameters in natural conditions (a pressure is more than 1 million atmospheres and temperature 10000^0C) do not allow us to synthesize diamonds in the conditions identical to natural ones. There is no material, which usable at such pressures and temperatures. Actually, the perspectives of development of relevant synthesis methods of diamond synthesis are restricted by pressure and temperature values. A technology of diamond synthesis was developed that aimed to remove these restrictions by applying neutron irradiation [5].

Energetic conditions necessary for graphite transition into diamond are created as a result of using explosion energy and neutron irradiation.

EXPERIMENT

Investigations were carried out using mixtures of manganese-nickel-graphite and copper-graphite. Graphite content in mixtures was equal to 8-10 weight percents. Spectroscopically pure graphite with particles size-50μm and powders of copper, nickel and manganese with particles size-8-50μm were used.

At the beginning loading of initial system metal-carbon in cylindrical ampoules, than explosion is stimulated by weak shock waves ~1.75gPa. Graphite is transformed into active state as a result of this processing. The next operation is an irradiation by fast neutrons in low-temperature regime (100-300K). After irradiation, samples were kept in a protection cell for to reduce radioactivity.

Irradiation energy is accumulated in continuous regime of reactor work (~100 hours) and conditioned by neutrons flow in a channel. Neutron dose is equal to ~$3*10^{22}$ N/m^2, energy E≥0.1meV; neutrons velocity is equal to $4.3*10^3$ km/sec.

Synthesized diamonds were extracted from reaction mixtures by chemical enrichment. Shock-wave loading leads to broadening of X-ray reflections peaks for all elements entered in reaction mixture. By analysis of X-ray lines profile it was found out that a broadenings is mainly connected to microstresses in samples at deformation. Irradiation leads to partial removal of microstresses. In this respect, irradiation is equivalent to “annealing”.

Shock action and after-neutron irradiation significantly change profiles of X-ray lines of materials. The results of measurement of X-ray reflection are presented in table 1.

Table 1. Alteration of intensities of X-ray reflections I/I_0

Condition	Mixture Mn-Ni-C			Mixture Cu-C		
	Ni(111)	C(00.2)	C(00.4)	Cu(200)	C(00.2)	C(00.4)
Initial sample	1,0	1,0	1,0	1,0	1,0	1,0
Explosion T_{exp} =300K	0,43	0,88	0,70	0,84	0,90	0,85
Irradiation-Explosion T_{ir} =300K	0,37	0,91	0,80	0,89	0,95	0,92

Static atoms dislocation is occurred as a result of dynamic forces, which leads to decreasing of integral intensity, a relative variation are calculated by the following formula:

$$\frac{I}{I_0} = e^{-2M}$$

where,

$$M = \frac{8\pi^2}{3} * \bar{U}^2 \sin^2 \vartheta / \lambda^2$$

$\bar{U}^2$- Mean-square displacement of static atoms
ϑ- Diffraction angle
λ- Wavelength of radiation

I/I_0 values was calculated for diffraction lines (00.2) and graphite (00.4), (200) and (111) nickel and (200) copper (Table 2 and figure 1) by using of given formulas.

Table 2. Dependence of relative integral intensities on displacement of static atoms

Substance	UHKL	U=0,05	U=0,10	U=0,125	U=0,20	U=0,30	U=0,50
graphite	(00.2)	0,997	0,989	0,982	0,955	0,903	0,752
	(00.4)	0,989	0,955	0,931	0,831	0,664	0,320
Cu	(200)	0,990	0,961	0,939	0,851	0,696	0,366
Ni	(200)	0,989	0,959	0,937	0,846	0,687	0,353
	(111)	0,992	0,969	0,952	0,882	0,755	0,457

A comparison of experimental and theoretical data showed that static atoms displacement in graphite after shock loading was equal to 20-30% of interplanar spacing in graphite. As a result of these dislocations, fluctuating minor areas can be appeared, in which a distance between graphite planes is less than in an initial structure. In these regions, favorable thermodynamic conditions are created by neutron irradiation action led to structural graphite reorganization in diamond.

A comparison of experimental and theoretical data was shown that a static dislocation in graphite after shock loading was equal to 20-30% of interplanar spacing in graphite. As a result of these displacements, fluctuating minor areas appear, in which a distance between graphite planes is less than in an initial structure. In these regions, favorable thermodynamic conditions are created by neutron irradiation leading to structural reorganization of graphite into diamond.

Interpretation of electron diffraction pattern obtained from irradiated samples showed that investigated structure corresponds to diamond. Characteristic feature of obtained diamonds is highly dispersed polycrystalline structure. A growth of diamond polycrystal has a stepped character. It can be explained by non homogeneous distribution of pressure during the irradiation process. It was also found well-faceted monocrystals of diamonds in samples (figure 2).

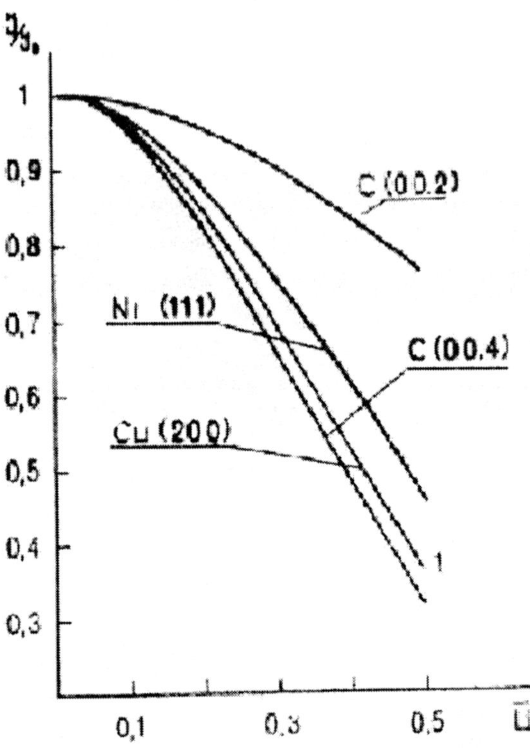

Figure 1. Dependence of relative integral intensities on static atoms dislocation.

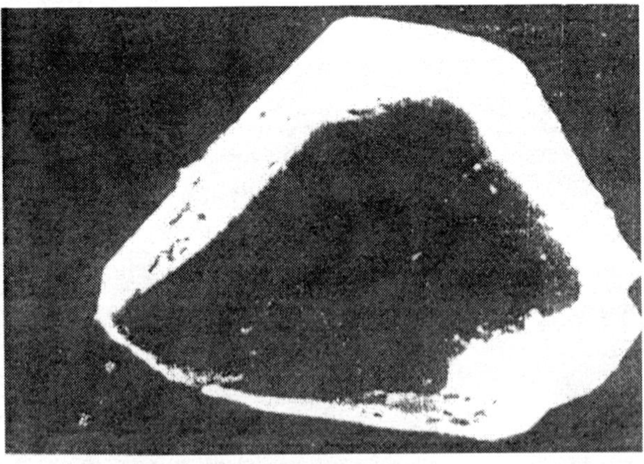

Figure 2. Electronic microphotography of diamond monocrystal.

Table 3. Theoretical and Experimental values of interplanar spacing of graphite and diamond

Interplanar spacing d					
Literature data nm				Experimental value nm	
diamond		graphite		diamond	
d	HKL	d	HKL	d	HKL
0,205	111	0,338	0,02	0,2067	111
0,126	220	0,212	1,00	0,1266	220
0,1072	311	0,202	1,01	0,1079	311
0,0885	400	0,169	0,04	0,0893	400
0,0813	331	0,123	11,0	0,0819	331

As known that deviation from idea crystal structure leads to essential changes of X-ray reflections spectra. In one case, a width of X-ray line is remained constant and an integral intensity is changed, that is formation of isolated particles of new phase takes place. In another case, a broadening is observed without any changes of integral intensity of lines. Such alterations are connected with linear dislocations and etc.

Obtained data show that a width of graphite line is remained constant in the range of experimental error, whereas, an integral intensity and lattice parameter are changed. Indicated alteration in a line profile can apparently be connected with a formation of inclusions of new phase. The presence of characteristic lines of metal carbides in X-ray pattern taken after irradiation is not observed. Therefore, the alteration in a line profile from graphite plane is connected to the formation of diamond inclusions. The table 3 bellow testifies this fact.

For implementation of phase transition of graphite into diamond it is necessary to change position of atoms in crystal lattice of graphite. Displacement of atoms from lattice site can cause decreasing of integral intensity and this is clearly observed in our experiments. A reason of indicated displacement can be formation of temperature gradient, which is formed in the process of irradiation by fast neutrons [6]. The bombarding particles transfer energy Q to lattice atom. This energy releases in the form of thermal energy in a small environment, and afterwards, distributes according to law of thermal conductivity. In this region atoms become strongly excited. Such stimulations are called temperature wedges. Typical energy values transferred to bombarded atoms are in the range from 10 to some hundreds keV. Thereby, as a typical wedge source can consider a knocked-on atom having defined energy. In the first approximation we can consider this atom as a source of spherical wedge with excitation energy Q. More accurately can be considered that excitation occurs as a consequence of superposition of wedges of less sizes located at different sites along the particle way.

Data of temperature distribution in graphite (in a thermal wedge) is presented (table 4), obtained by solution of thermal conductivity equation [7]. Temperature distribution has a maximum in the centre and quickly drops with the increasing of distance.

The change of mechanical properties at irradiation is mainly caused by changing of aggregative state at local sites. Alteration of specific volume at local sites leads to appearance of stress and formation of zones of plastic deformation; its radius is determined by method given in the article [8]. For graphite the radius of plastic deformation is equal to $r \approx (1.72 \div 5.51) * 10$nm.

Appeared stresses around a thermal wedge can be relaxed by plastic deformation, if a matrix surrounded it is flexible. In the case when matrix is not plastic and friable, cracks are formed around the thermal wedge. In this case it should be appeared a noticeable broadening of X-ray reflections, which was not observed in our experiments. We can consider that mentioned stress are relaxed by phase transition around extremal zone.

Table 4. Dependence of temperature distribution in a thermal wedge on energy of incident neutron

Energy of incident neutron,E, mev	Distance from wedge centre r , nm	Temperature in a thermal wedge, T(r,t)
0,1	1	$4{,}3 \cdot 10^3$
	10	$3 \cdot 10^3$
	50	0,58
1,0	1	$4{,}3 \cdot 10^4$
	10	$3 \cdot 10^4$
	50	5,8
3,0	1	$13 \cdot 10^4$
	10	$9 \cdot 10^4$
	50	17,4

For confirmation of obtained results irradiation of pure graphite preliminary processed by shock waves and without it was carried out. As it was seen, graphite irradiation without metal in reactor does not cause essential alterations. A temperature decrease in wedge occurs for 10^{-10}-10^{-11}sec. Stresses appeared in the process of formation of temperature gradient are removed without plastic deformation. A phase transition is not observed.

Irradiation by fast neutrons is a nodal moment of diamond synthesis in the developed technology. Reactor working for only diamond synthesis is not beneficial from economical point of view. Therefore, it is reasonable to use active reactors for this purpose. Such an approach to the assigned task gives us the possibility to increase efficiency of developed technology of diamonds synthesis.

REFERENCES

[1] B.V. Nekrasov. "Basis of general chemistry", *"Chemistry"*, volume 1, pp.656, 1973.

[2] Z.M. Galimov. "About evolution of Earth carbon", "*Geochemistry*", 15, 1967.

[3] A.V.Uxanov, T.V.Malisheva, "*Geochemistry*", 10, 1973.

[4] Y.A. Kalashnikov, "Physical chemistry of substances at high pressures", "*High school*", 239, 1987.

[5] A.D. Nozadze, T.Sh.Pipia, E.Sh.Chagelishvili, I.V. Chxartishvili, T.G. Mardaleishvili-patent Georgia #369, "*Obtaining method of synthetic diamonds*", V 21 D 22/00, C 01 V 31/06, 19.09.1996.

[6] M.P. Shaskolskaia, "Crystallography", "*High school*", 391.

[7] B.N. Yudaev, "Heat transfer", "*High school*", 360, 1973.

[8] I.M.Lifshic et all., " To theory of irradiated changes in metal", "*Atomic energy*", N6, 1969.

In: New Developments in Material Science ISBN 978-1-61668-852-3
Editors: E. Chikoidze and T. Tchelidze

Chapter 16

EQUATORIAL KERR EFFECT IN ULTRAFINE MAGNETIC STRUCTURES

Lali Kalandadze*

Department of Physics, Batumi State University,
35 Ninoshvili street, Batumi, 6010, Georgia

ABSTRACT

In the present paper, using magnetite magnetic fluids as examples, we consider theoretically and experimentally the optical and magneto-optical properties of magnetic fluids based on particles of magnetic oxides, for which the relation $k^2 << n^2$ holds (n and k are the optical constants of the material).

In this work the equatorial Kerr effect is represented within the theoretical Maxwell-Garnett model. A theoretical analysis has shown that in the mentioned case of magnetic oxides, the equatorial Kerr effect $\delta_e(q)$ for magnetic fluids with the ratio of the volume occupied by magnetic particles is related to the equatorial Kerr effect δ_m on the material of particles by the simple relation.

The estimations showed that the amount of effect is proportional to the occupancy of the volume of the magnetic fluid with magnetic particles - q. Besides, the spectral dependences of magneto-optical effects are similar for the magnetic fluids and for the magnetic particles.

* E-mail: lali62@mail.ru

The experimental and theoretical results for the magnetite magnetic fluids are in a good agreement.

The result is expected to be applicable to all the magnetic ultrafine medium, the optical constants of which, n and k, holds this relation $k^2 << n^2$.

1. Introduction

Magneto-optical research methods are successfully used to solve different types of task in the physics of magnetic phenomena. They are the powerful tools means to observe domains and the boundaries of domains, as well as to examine the electronic energy structure of the solids [1]. Nowadays, magneto-optical research methods are widely used for examination of ultrafine magnetic structures with insertions that have the sizes of structural and magnetic heterogeneities less than 1000 $\overset{0}{A}$. Magneto-optical investigation of these types of structures are subject of overall interest, which is conditioned by both theoretical and practical importance. The scientific interest is explained by the unique properties of ultrafine magnetic structures compared to the solid ones. For instance, the new effects were discovered when studying magneto-optical properties of ultra-fine structures: equatorial Kerr effect production on the s-component of the incident light, the breach of linear dependences between magnetization and magneto-optical effects, the increase of equatorial Kerr effect in ultrafine structures compared to bulk materials, absence in bulk ferromagnetic [2]. It has been discovered in the course of magneto--optical investigation of utrafine magnetic structures that they are exception from to the classic rules of physics. This complicates the task of interpretation of the existing experimental data, as well as establishing of the appropriate laws.

On the other hand, practical interest is determined by the widespread application of the ultrafine magnetic structures in industry and in everyday - life: magnetic fluids, thin discontinuous metal films, ferrite-garnet films, implanted magnetic surfaces.

2. Theory

In general magneto-optical effects can be divided into two groups: 1) Faraday and Cotton-Mutton Effects produced by the light which propagates through magnetic crystal, and 2) Magneto-Optical Kerr Effects which are created by the light reflected from the magnetic crystal.

The magneto-optical reflection effects are successfully used to study the processes in the surface regions of magnetic fluids. By these methods it has been revealed that the application of an external magnetic field to the magnetic fluids causes changes in the concentration of magnetic particles in the surface region of magnetic fluids, creates the surface layers, magnetized apposite to the external magnetic field, stimulates the surface lamination of the magnetic fluids [3].

When considering the magneto-optical effects, which are odd function of magnetization, in magnetic fluids, it should be kept in mind that magnetic fluid is a variety of the ultrafine media [4].

The magneto-optical properties of ultrafine media were analyzed in refs. [2,5] It was shown that when considering the magneto-optical properties of medium consisting of magnetic colloidal particles the sizes of which are much less than the light wavelength, one should introduce the tensor of the effective dielectric permittivity

$$\varepsilon_{\mathrm{eff}} = \begin{bmatrix} \varepsilon_{\mathrm{eff}} & -i\varepsilon'_{\mathrm{eff}} & 0 \\ i\varepsilon'_{\mathrm{eff}} & \varepsilon_{\mathrm{eff}} & 0 \\ 0 & 0 & \varepsilon_{0\mathrm{eff}} \end{bmatrix}, \quad (1)$$

where, $\varepsilon_{eff} = \varepsilon_{1eff} - i\varepsilon_{2eff}$, $\varepsilon_{0eff} = \varepsilon_{01eff} - i\varepsilon_{02eff}$ and $\varepsilon'_{eff} = \varepsilon'_{1eff} - i\varepsilon'_{2eff}$.

In this case the tensor components depend on both the properties of magnetic colloidal particles themselves and the properties of the medium in which they find themselves.

Diagonal component of (1) is connected to reflective index n_{eff} and absorption index k_{eff} of ultrafine media by formula

$$\varepsilon_{eff} = (n_{eff} - ik_{eff}) \quad (2)$$

For magnetic fluids with a low concentration of magnetic colloidal particles and consequently, with no interaction between them, the tensor components of the effective dielectric permittivity within the framework of the theoretical Maxwell-Garnett model of an effective medium, can be written as:

$$\varepsilon_{eff} = \frac{2q(\varepsilon_m - \varepsilon_0) + (\varepsilon_m + 2\varepsilon_0)}{(\varepsilon_m + 2\varepsilon_0) - q(\varepsilon_m - \varepsilon_0)}\ ;\ \varepsilon'_{\text{eff}} = \frac{9q\varepsilon_0^2\varepsilon'_m}{(\varepsilon_m(1-q) + \varepsilon_0(2+q))^2} \quad (3)$$

where $\varepsilon_m = \varepsilon_{1m} - i\varepsilon_{2m}$ and $\varepsilon'_m = \varepsilon'_{1m} - i\varepsilon'_{2m}$ are the diagonal and nondiagonal tensor components of the dielectric permittivity of the material of magnetic colloidal particles, ε_0 is the dielectric permittivity of the fluid phase, and q is the ratio of the volume, occupied by magnetic particles, to the total volume of the magnetic fluid.

The equatorial Kerr effect is related to the tensor components of the effective dielectric permittivity as follows

$$\delta_e = \frac{2\sin 2\varphi(A\varepsilon'_{1eff} + B\varepsilon'_{2eff})}{A^2 + B^2} \quad (4)$$

where $A = \varepsilon_{2eff}(2\varepsilon'_{1eff}\cos^2\varphi - 1)$ and $B = (\varepsilon_{2eff}^2 - \varepsilon_{1eff}^2)\cos^2\varphi + \varepsilon_{1eff} - \sin^2\varphi$, where is the angle of light incidence on the sample.

In general, the magneto-optical properties of magnetic fluids are very different from the properties of the bulk ferromagnetic. This fact causes the changes in spectral and angular dependences of the magneto-optical effects. The reason for this is that magnetic fluids represent an example of a magnetic ultrafine medium composed with magnetic particles, the sizes of which are much less than the light wavelength. Therefore, it is important to define clearly the conditions of the magneto-optical experiment to make easier the interpretation of the magneto-optical spectrums, which in its turn, provides with the information of surface layer of magnetized magnetic fluids, electronic energy structure of fine magnetic particles, as well as of the properties of carrier fluids. Besides, it is significant to examine the electronic energy structures of fine magnetic particles not only to investigate the temporary stability of magnetic fluids, but also to analyze the chemical processes on the surface of the magnetic particles. On the other hand, in some cases the fine magnetic particles could be used as a probe.

In the present paper, using magnetite magnetic fluid as an example, we consider theoretically and experimentally the optical and magneto-optical properties of magnetic fluids based on particles of magnetic oxides, for which the relation $k^2 << n^2$ holds; n and k are optical constants of the material.

The investigation of the equatorial Kerr effect is represented within the framework of the theoretical Maxwell-Garnett model.

In this case, considering the following relation $k^2 << n^2$ and formula (2), A and B in the formula (4) can be expressed in this way:

$$A = 2n_{eff}k_{eff}(2n_{eff}^2\cos^2\varphi - 1)\,;\; B = (n_{eff}^2 - 1)(1 - (n_{eff}^2 + 1)\cos^2\varphi) \quad (5)$$

The analysis of these expression shows that in definite range of the angles of light incidence φ (for magnetite particles with different q, φ=69-72^0), $A << B$, which allows us to rewrite formula (4) in the simpler form:

$$\delta_e = \frac{2\sin 2\varphi}{B}\varepsilon_{2eff}' \quad (6)$$

Imaginary part of nondiagonal component ε_{2eff}' of the tensor of the effective dielectric permittivity ε_{eff}' (3) can be written down in following way:

$$\varepsilon_{2eff}' = q\varepsilon_0^2\left[\frac{\varepsilon_{2m}}{(Y(\varepsilon_{1m} - \varepsilon_0) + \varepsilon_0)^2} - \frac{2Y\varepsilon_{2m}\varepsilon_{1m}'}{(Y(\varepsilon_{1m} - \varepsilon_0) + \varepsilon_0)^3}\right] \quad (7)$$

where, $Y = \dfrac{1-q}{3}$ (8)

Taking into account that dielectric permittivity for isotropic and transparent surrounding area is $\varepsilon_0 = n_0^2$, formula (6) takes the form:

$$\varepsilon_{2eff}' = qn_0^4\left[\frac{\varepsilon_{2m}'}{(Y(n_m^2 - n_0^2) + n_0^2)^2} - \frac{4Yk_m n_m \varepsilon_{1m}'}{(Y(n_m^2 - n_0^2)^3}\right] \quad (9)$$

where n_m and n_0 are the reflective indexes of particles and liquid carrier accordingly.

If $n_0 \approx n_m$, then

$$\varepsilon'_{2eff} = qn_m^4\left[\frac{\varepsilon'_{2m}}{n_m^4} - \frac{4Yk_m\varepsilon'_{1m}}{n_m^5}\right] \tag{10}$$

the formula (7) shows that Y ranges between 0 - 0,3 and is less or equal to k_m This leads to the fact that previously discussed case when $k^2 << n^2$, it is a possible to ignore the second term in the right part of (10)and write down the expression for the equatorial Kerr effect in the simpler way:

$$\delta_e(q) = q\delta_m \tag{11}$$

where, $\delta_e(q)$ is equatorial Kerr effect for magnetic fluids with the ratio of the volume occupied by magnetic particles *q;* δ_m - equatorial Kerr effect of the material of particles.

It is necessary to say that obtained results (11) takes place if only these three conditions are followed: 1) $k^2 << n^2$; 2) angle of light incidence φ fulfills the requirement $A << B$; 3) $n_0 \approx n_m$.

It follows from the relation obtained that in the specific experimental conditions the amount of equatorial Kerr effect is proportional to q, and therefore to magnetization itself. Also, the character of spectral dependences of magneto-optical effects are similar for the magnetic fluids and the substance of particles.

3. Experimental

For experimental investigation of magnetic fluids we chose the equatorial Kerr effect. The equatorial Kerr effect consists in a change in the intensity of linearly polarized light reflected from the sample in the case of reversal of magnetization. It can be written as:

$$\delta = \frac{I_H - I_{H=0}}{I_{H=0}}, \tag{12}$$

where, I_H and $I_{H=0}$ are, respectively, the intensities of light reflected from the magnetized and demagnetized sample.

The optical constants were determined using the Every method [6, 7].

We have investigated magnetite magnetic fluids with different carrier fluids (water, kerosene, silicon-organic compound) and its sediments.

RESULTS AND DISCUSSION

Figure 1 represents the results of calculations of $\delta_e(q)$ made by means of the formula (4) for magnetite particles in the air (a), in the kerosene (b) and in the liquid silicon-organic compound (c), and by the formula (11). Calculations were carrier out for the fixed angle of light incidence and $\varphi = 70^0$ different q. From the Figure 1 it is obvious that the magneto-optical maxima for magnetite particles shift to the side of large energies. If we increase ε_0, this shift diminishes, and for the liquid silicon-organic compound ($\varepsilon_0 = 2,56$) the shift approximates to the zero. The results derived from our research indicat that calculations conducted according to (4) for φ=69-72^0 and large ε_0 ($\varepsilon_0 = 2,56$), coincide with the calculations carried out by the simplified formula (11). The experimental results obtained for the most of the magnetite fluids investigated are in a good agreement with the calculations made by the formula (11). Figure 2 gives the experimental dependences of the equatorial Kerr effect on the incident light photon energy $\hbar\omega$ for magnetic fluids based on magnetite particles in the liquid silicon-organic compound, for its sediments and for a bulk film Fe_3O_4. In the course of investigation of the equatorial Kerr effect at two angles of light incidence and of the optical constants, we determined the spectral dependences of the really ε'_{1m} and imaginary ε'_{2m} parts of the nondiagonal components ε'_m of the tensor of dielectric permittivity for the bulk magnetite to checking the correlation between (5) and (10) which reflects a linear dependence between $\delta_e(q)$ and ε'_{2m}.

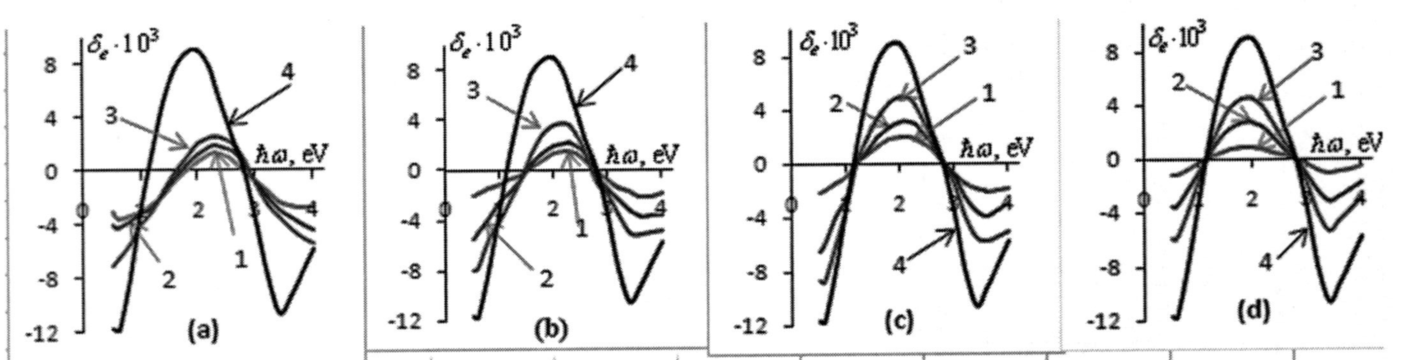

Figure 1. Dependences of the equatorial Kerr effect on the photon energy of incident light $\hbar\omega$, calculated by formula(4) for magnetite particles in the air (a), in the kerosene (b) and in the liquid silicon-organic compound (c), and by the formula (11) (d) with q =0,1(1); 0,3(2); 0,5(3); 1(4).

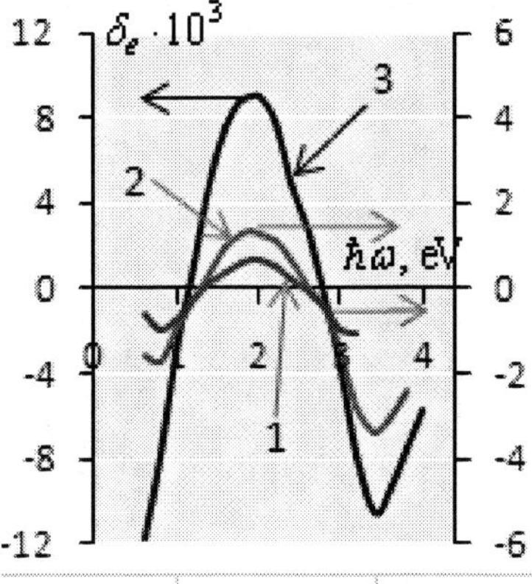

Figure 2. Experimental dependences of the equatorial Kerr effect on $\hbar\omega$ for magnetic fluids based on magnetite particles in the liquid silicon-organic compound (1), for its sediments (2) and for bulk film Fe_3O_4 (3).

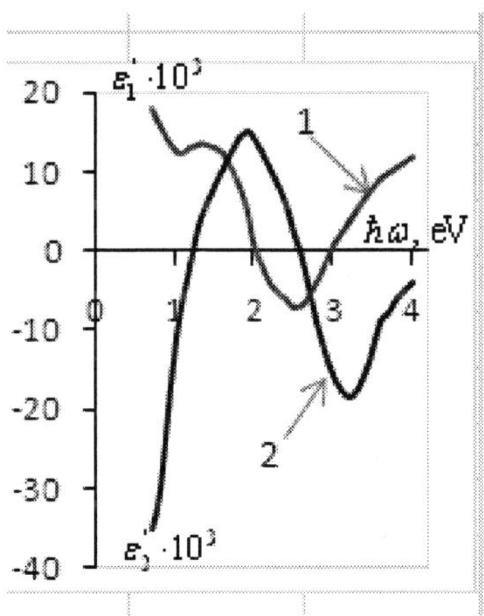

Figure 3. Experimental dependences of the equatorial Kerr effect on $\hbar\omega$ for magnetic fluids based on magnetite particles in the liquid silicon-organic compound (1), for its sediments (2) and for bulk film Fe_3O_4 (3).

If compare illustrated frequency dependences of the equatorial Kerr effect (Figure 1) with the dependences of $\varepsilon_{1m}^{'}$ and $\varepsilon_{2m}^{'}$ (Figure 3), it is obvious that the dependences of the equatorial Kerr effect and the imaginary part of the nondiagonal component of the tensor of dielectric permittivity of magnetite on incident photon energy are in a good correlation, which also testifies (5) and (10) formulations.

Conclusion

In the present paper, on the example of magnetite magnetic fluids we considered theoretically and experimentally the optical and magneto-optical properties of magnetic fluids based on particles of magnetic oxides. Within the framework of the theoretical Maxwell-Garnett model the simple relation between the equatorial Kerr effect for ultra-fine medium and bulk material is revealed. The conditions of the magneto-optical experiment were defined for the essential analytical simplification of the subsequently received results. Moreover, it clearly was illustrated that the experimentally derived data are well compatible with the above discussed theory.

In conclusion, we would like to underline that this result are expected to be suitable to all magnetic ultrafine medium with the same relation between optical constants: $k^2 << n^2$.

References

[1] Krinchik G.S. *Physical Principles of Magnetic Phenomena*, Moscow State University 1985, pp.294-330.

[2] Nikitin L V and Kasatkina O V 1986 Proc. X All-Union school-seminar on New magnetic materials of microelectronics (Riga*)* vol 2 124 (in Russian).

[3] Nikitin L V and Tulinov A A 1990 *Journal of Magnetism and Magnetic Materials* 85 89.

[4] Nikitin L V, Kalandadze L G, Akhmedov M Z, Nepijko S A 1995 *Journal Magnetism and Magnetic Materials* 148 279.

[5] Krinchik G.S. Nikitin L.V. and kasatkina O.V. Poverkhnost', Fizika, Khimiya, *Mekhanika* 7 (1985) 140.

[6] Avery D. *Proc. Phys. Soc. Of London* (B), 1952, v.65, 426-429.

[7] Sokolov A. V. *Optical properties of metals*; M, 1961.

In: New Developments in Material Science ISBN 978-1-61668-852-3
Editors: E. Chikoidze and T. Tchelidze

Chapter 17

CELLULOSIC MATERIALS PROCESSING BY CHILLING AND MECHANICAL MILLING METHOD

T. B. Dudauri[1], M. T. Chachkhiani[1], M. G. Berezhiani[1] G.Sh. Partskhaladze[1], L. N. Tsiklauri[1], V. D. Ugrekhelidze[1], N. G. Zakariashvili[2] and A. M. Berejiani[*3]

[1]National High Technology Center of Georgia,
[2]Durmishidze Institute of Plant Biochemistry and Biotechnology- Georgian Academy of Science,
[3]Georgian Technical University

ABSTRACT

Scientific investigations show that hydrolysis rate of native cellulosic compounds under enzymatic fermentation is low and therefore pretreatment of cellulosic wastes becomes one of the topical points. Cellulose pretreatment method based on combination of chilling and mechanical milling of cellulosic wastes was used for investigation of structural changes of different cellulosic materials. When frozen water contained in cellulosic wastes expands it causes the mechanical stress that

---

* E-mail: a.berezhiani@soon.com

is followed by partial destruction of crystalline cellulose. Cellulose decomposition rate, changes of both: cellulose polymerization degree and specific surface area have been determined when cellulosic wastes such as: sawdust, cardboard, newsprint, filter paper and composite of all above materials were frozen at 0, -20, -70 and -196^{0}C. Studies showed that mechanical milling of frozen cellulosic wastes enhances subsequent breaking of cellulose that leads to further increase of cellulose specific reactionary surface accessible for enzymes. Investigations conducted under the proposed research showed that maximum increase of specific surface area of cellulosic wastes could be reached when cellulosic wastes where chilled at -20^{0}C and milled at once.

Biomass encompasses a range of products derived from photosynthesis and is essential chemical storage of solar energy. In developing countries it is basically used as a source of thermal energy. On the other hand different forms of biomass that include fuel wood, energy crops (crops specifically grown for energy), agricultural and forestry residues, food and timber processing residues, municipal solid wastes, sewage etc., can be converted into various carbon-based fuels such as: biogas, ethanol and methanol and thus contribute in partial replacing of fossil fuels in the energy industry. Among the biomass-to- energy conversion technologies biological methods are considered more environmentally friendly ones and improvement of biomass fermentation process becomes one of the topical points for new developments.

Scientific investigations show that hydrolysis rate of native cellulosic compounds under enzymatic fermentation is low and therefore pretreatment of cellulosic wastes becomes one of the topical points.

Pretreatment of cellulosic biomass in a cost-effective manner is major challenge of cellulose to bio-fuel technology research and development. Developing technologies that may decrease the cost of bio-fuel production from different composition cellulosic wastes can be considered in terms of pretreatment, fermentation, alcohol recovery, by-product recovery, and waste treatment.

Cellulose pretreatment method based on combination of chilling and mechanical milling of cellulosic wastes was used for investigation of structural changes of different cellulosic materials.

Among the diverse spectrum of biomass, cellulosic wastes are the most abundant natural organic compounds on the earth and are in general the least expensive carbohydrate source. This is why these wastes attract much attention for bio-fuel (ethanol or biogas) production. In general the types of natural cellulose could be divided in two main species: lignocelluloses such as: wood, shrubs, leaves, herbs, algae and pure cellulose: cotton and flax. Both species of cellulose mainly have crystalline structure that makes complicate its biodegradation. Besides that, lignin that is one of the components of lignocelluloses prevents enzymes' accessibility to cellulose.

Many investigations show that reactivity of cellulosic compounds under enzymatic fermentation is low and therefore pretreatment of cellulosic wastes becomes one of the topical points. Pretreatment refers to the solubilization and separation of one or more of the four major components of biomass – hemicellulose, cellulose, lignin and extractives – to make the remaining solid biomass more accessible to further biological or chemical treatment. Hydrolysis (saccharification) breaks down the hydrogen bonds in the hemicellulose and cellulose fractions into their sugar components: pentoses and hexoses. These sugars can then be fermented into ethanol that commonly is used as a fuel and/or food substance. Apart from this, ethanol is one of the precursors for methane formation when digesting organic wastes under anaerobic conditions.

Various pretreatment methods, such as physical, chemical and biological ones are described in scientific publications which promote the accessibility of polysaccharides in a lignocellulose complex for enzymatic hydrolysis. Physical treatment breaks down the feedstock size by milling or aqueous/steam processing. Following to literature data, steam explosion method is considered as the most effective physical method for cellulose pretreatment as it leads to cellulose defibration. The most common chemical methods used for cellulosic feedstock are mild and concentrated acid, alkaline, organic solvent, ammonia, sulfur dioxide, carbon dioxide or other chemicals to make the biomass more digestible by the enzymes. (1-4).These methods often involve such conditions that lead to the formation of degradation products which act as inhibitors for enzymatic hydrolysis (5). In general each type of feedstock requires combination of pretreatment methods to optimize the yields of that feedstock and maximize the sugar yield.

Cellulose pretreatment method first introduced here is based on combination of chilling and mechanical milling of cellulosic wastes. The idea lies in application of anomalous characteristic of water. The point is that when water temperature is decreased bellow $+4^0$ C the water expansion factor's value becomes negative: $\alpha<0$. On phase transfer: water-ice, specific volume of water increases by 10% (6). Under further decrease of ice temperature anomalous specific volume dependence upon the temperature remains almost the same. When frozen water contained in cellulosic wastes expands it causes the mechanical stress that is followed by partial destruction of crystalline cellulose.

At the initial stage of experiments cellulose and water percentages were determined in the following cellulosic wastes: sawdust, newsprint, cardboard and filter paper. Polymerization degree and specific surface area of non treated sawdust, newsprint, cardboard and filter paper have been determined. Analytical samples had been frozen at 0^0 C, -20^0 C, -70^0 C and -196^0 C and kept at these temperatures during 2 hours. One set of frozen samples have been thawed and afterwards milled. Other ones had been milled at once, in the frozen state. Cellulose hydrolysis rate, change of both: cellulose polymerization degree and cellulose specific surface area of each analytical sample had been determined after the treatment of above analytical samples.

Cellulose hydrolysis rate was determined by percentage reduction of initial cellulose in accordance with the standard Apdegraph method that was based on determination of optical density of preliminarily treated cellulosic substrate.

Results are given in table #1.

Experimental results show that maximum decomposition rate of sawdust and newsprint is reached at -196^0C, as for cardboard, filter paper and composite of all above cellulosic wastes, maximum reduction of initial cellulose is obtained at -70^0C when the analytical samples are milled in the frozen state.

Then, Cellulose polymerization degree has been determined by the viscosimetry method: Samples of cellulosic wastes were dissolved by coppery-ammonia complex and the viscosities of obtained solutions were measured by using of Ostwald viscosimeter. Degree of polymerization was calculated by the following formula:

$$P= 2000\ \eta_k /(C \times (1+0,29\ \eta_k))$$

where C is concentration of cellulose (gr/l), η_k is the specific viscosity that equals to: $(\tau / \tau_0) - 1$, where τ - is flow down time of the treated samples, τ_0 is flow down time (sec) of pure dissolvent. Results are given in table 2.

Data given in table 2 indicate that polymerization degree of filter paper, cardboard and composite of above cellulosic wastes decrease when they are frozen at – 196 ^{0}C and milled at once but newsprint's polymerization degree lessens when it is frozen at -20^0C.

Table 1

Cellulosic substrate	Chilling temperature							
	-3^0 C		-20^0 C		-70^0 C		-196^0 C	
	milled in frozen state	milled after thawing	milled in frozen state	milled after thawing	milled in frozen state	milled after thawing	milled in frozen state	milled after thawing
	Reduction of initial cellulose (% weight)							
Sawdust	2,43	4,42	4,63	4,50	4,90	4,82	5,56	3,86
Newsprint	0,81	1,80	1,0	1,8	3,85	3,50	5,05	3,57
Cardboard	3,97	5,38	3,81	3,92	6,29	5,73	4,31	5,33
Filter paper	1,98	2,32	2,46	2,60	6,87	2,54	4,60	2,89
Composite	3,24	3,34	3,31	3,65	9,25	3,92	8,46	7,33

Table 2

Cellulosic wastes	Polymerization degree				
		Chilling temperature ^{0}C			
	non treated sample	0	-20	-70	-196
newsprint	1190	733	301	411	360
filter paper	522	461	384	268	244
cardboard	713	687	464	270	252
composite	670	544	372	260	152

Afterwards, cellulose specific surface area has been determined. For definition of cellulose specific surface area the method based on chemotripsin absorption was used.

Calculations have been done by using the following equation:

$$S=S_1 \times a \times N_a / C \ (m^2/gr)$$

where $S_1=10^{-17}$, a - is the amount of absorbed ferment (mol/l), C -is the concentration of cellulose (g/l), N_a- is the Avogadro's number.

Results are given in table #3.

Results obtained show that for all analytical samples maximum increase of specific surface area is achieved when above cellulosic wastes are frozen at -20^{0}C and milled in the frozen state.

The mechanical milling of frozen cellulosic wastes enhances subsequent breaking of cellulose that leads to further increase of cellulose specific reactionary surface accessible for enzymes. In addition laboratory experiments showed that frozen cellulosic wastes were reduced to smaller particles in comparison with milling of wet or dry ones.

Taking into consideration the fact that specific surface area is the determinative factor for cellulose fermentation by using of fungi or clostridia species, chilling at -20^{0}C with further milling of frozen cellulosic wastes could be considered one of the promising pretreatment methods.

Table 3.

Cellulosic wastes	Specific Surface Area (m^2/gr)			
		Chilling temperature ^{0}C		
	non treated sample	-20	-70	-196
newsprint	0,016	0,036	0,014	0,028
filter paper	0,038	0,05	0,038	0,38
cardboard	0	0,037	0,031	0,018
composite	0,002	0,032	0,026	0,017

REFERENCES

[1] Fan L.T. et all. The nature of lignocellulosics and their pretreatment for enzymatic hydrolysis. *Adv. Biochem. Eng. Biotechnol.* 1982, 23:1 58-87.

[2] Parisi F. Advances in lignocellulosics and in the utilization of the hydrolyzates, *Biochem. Eng. Biotechnol.* 1983, 38: 54-87.

[3] Bjerre A.B. Olesen A.B. et all. Pretreatment of wheat straw using wet oxidation and alkaline hydrolysis resulting in convertible cellulose and hemicellulose. *Biotechn. Bioeng.*, 1996, 49:568-577.

[4] Zheng Y, Lin H-M et all. Supercritical carbon dioxide explosion as a pretreatment for cellulose hydrolysis. *Biotechn. Bioeng.* 1995, 17:845-850.

[5] Olsson L. Fermentation of lignocellulosics hydrolyzate for ethanol production. *Enzime Microb. Techn.*, 1996, 18:312-331.

[6] Spravochnik po fizikoteknicheskih osnovah kriogeniki, Moskva, *Energoatomizdat*, 1985.

In: New Developments in Material Science ISBN 978-1-61668-852-3
Editors: E. Chikoidze and T. Tchelidze

Chapter 18

DEPOSITION OF SIO_2 FILMS FROM TETRAETHOXYSILANE-BASED SOLUTION BY THE SOL-GEL METHOD

***T. I. Pavliashvili**[*1], M. V. Janjalia[1] and M. I. Gudavadze[2]*

[1]Andronikashvili Institute of Physics, Tamarashvili str.N6, Tbilisi 0179, Georgia

[2]Javakhishvili State University of Tbilisi, Ccavchavadze av.N,1 Tbilisi 0128, Georgia

ABSTRACT

Deposition of Sio_2 films was made from the solution prepared on the basis of tetraethoxysilane. The “maturing” process of film-forming solution was controlled by high-resolution nuclear magnetic resonance (nmr) method (on hydrogen nuclei) and, at the same time, was checked by testing depositions. The deposition of films was carried out by the method of centrifugation on “кдб-10”- grade silicon plates having the hole conductivity and <111> crystallographic orientation. For determination of the composition of films we studied infra-red (IR) spectra registered on “UR-20” spectrometer.

On the basis of the obtained films, the metal-dielectric-semiconductor Al-SiO_2-Si-Al structures were prepared. Some electro-physical characteristics have been studied: electric strength, dielectric

[*] Email: medeadodo@gmail.com

permeability and dielectric losses. The measurement of these characteristics revealed the prospects of the obtained films both, in microelectronics and in nanotechnologies.

Dielectric films find a wide application in microelectronic technology. Depending on the functions of films, they should meet different requirements. For films used as a inter-layer isolation the important characteristics are: the electric strength, the electrical specific resistance, the dielectric loss tangent (tg δ), the dielectric permeability, as well as the uniformity of thickness and the stability of these properties. In manufacturing of integral circuits (IC) by using the multi-level metallization, the number of commutation levels, the time of the delay of signal propagation and the power released in the system of inter-connections are increased. Besides, at the places of intersection of conductors, the dielectric film should produce the capacitance not higher than 2 pf. traditionally, in microelectronics, inorganic dielectrics are used as the dielectric films. Silicon dioxide still remains to be the most widely used dielectric deposited by different methods. Among these methods we can distinguish: high-temperature oxidation of silicon, pyrolysis, chemical deposition from gaseous phase (CVD) [1], chemical deposition from the solutions (CSD) [2, 3], plasma anodizing and plasma-chemical deposition.

Recently, the new technologies of deposition of dielectric films with low ($\varepsilon<3.8$) and ultra-low ($\varepsilon<2.2$) dielectric permeability have been actively developed. To obtain the dielectric films with low dielectric permeability, it is possible to use both inorganic and organic compounds, as well as the hybrid inorganic – organic compounds providing the dielectric permeability $\varepsilon=2.2\text{-}3.5$ [4].

Among the studied compounds used in microelectronic technology we can single out tetraethoxysilane. Film-forming compounds are usually used in the form of solution. Tetraethoxysilane can be used both, for obtaining SiO_2 dielectric films and for deposition of multi-component films of various compositions [5].

The reactions of hydrolysis and poly-condensation of the products of tetraethoxysilane hydrolysis make the basis for the process of deposition of SiO_2 films from tetraethoxysilane-based solution. The character of proceeding of these reactions depends on many technological factors. The requirements imposed on the solutions can be considerably satisfied by a correct choice of solvents. In the solutions of film-forming materials the optimal ratio should be provided between the initial basic film-forming material, the solvent and the catalyst. The used solvent should be mixed well with water. To obtain the high-quality films of necessary thickness, the film-forming solutions should have low viscosity and low surface tension providing a good wetting of semiconductor and metal surfaces. The necessary condition for formation of films is the fast process of hydrolysis in the thin layer. The solvent and the products of hydrolysis not included in the composition of film should not have a low vapor pressure at room temperature.

The important factors having the effect of the properties of film-forming solutions are the concentration of tetraethoxysilane in the solution and the molar ratio of $Si(OC_2H_5)_4/H_2O$, as well as the acidity (PH) medium. The nature of catalyst has a strong influence on the process of tetraethoxysilane hydrolysis. Hydrolysis in case of alkali catalysts leads to the formation of solid reaction products. Hydrochloric and nitric acids are often used as a catalyst [6].

EXPERIMENT

The work studies the process of deposition of SiO_2 films from the solution prepared on the basis of tetraethoxysilane. The optimal conditions of film-deposition are determined and some properties of the films are studied. To prepare the film-forming solution, the following preparations were used: tetraethoxysilane 18 ml, hydrochloric acid – 0.04 ml, acetone – 60 ml and water – 4ml.

The spectra of nuclear magnetic resonance of high resolution for hydrogen nuclei were registered on spectrometer “Jeol – C 60 –HL” at 60 MHz frequency. Figure 1 shows the proton resonance spectrum of tetraethoxysilane consisting of two multiplets at 0.5 ppm and 3.2 ppm, and Figure 2 – the proton resonance spectrum of film-forming solution on the basis of tetraethoxysilane. The spectra of nuclear magnetic resonance were registered for observing the evolution of the processes of solution “maturing” in time. At the same time, the testing deposition of films was made. The experiments showed that after 36 h, the solution gained the property of film-formation [7].

Deposition on silicon substrate of p-type conductivity, with electrical specific resistance ρ=10 om.cm of crystallographic <111> orientation and 60 mm diameter was carried out by the method of centrifugation at the velocity of 2000 revs/min.

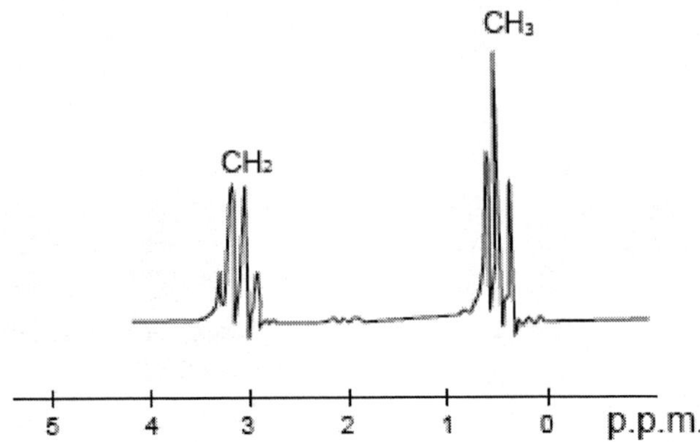

Figure 1. H^1 NMR Spectrum of Tetraethoxysilane.

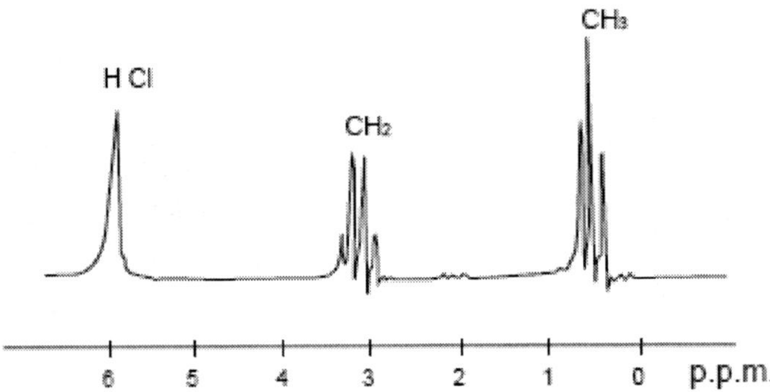

Figure 2. H^1 NMR Spectrum of film -forming Solution Based on Tetraethoxysilane.

After drying in thermostat at 130^0C during 30 min., the temperature of thermostat was increased gradually up to 190^0C and the films were kept for 40 min. As a result of such thermal effect, the final hydrolysis and the decomposition of intermediate products took place.

For a full removing of solvent and organic residues, of a main part of moisture, and for decomposition of acids, the obtained films were treated in diffusion furnace at the temperature of 470^0C for an hour. After the thermal annealing, rather durable films were formed with a good adhesion to the silicon substrate. The analysis of IR spectrum (Figure 3) registered on UR-20 spectrometer showed that the films had the absorption region in the range of 1000-1150cm^{-1} in correspondence with Si-O coupling. Other intermediate compounds and organic residues were not registered.

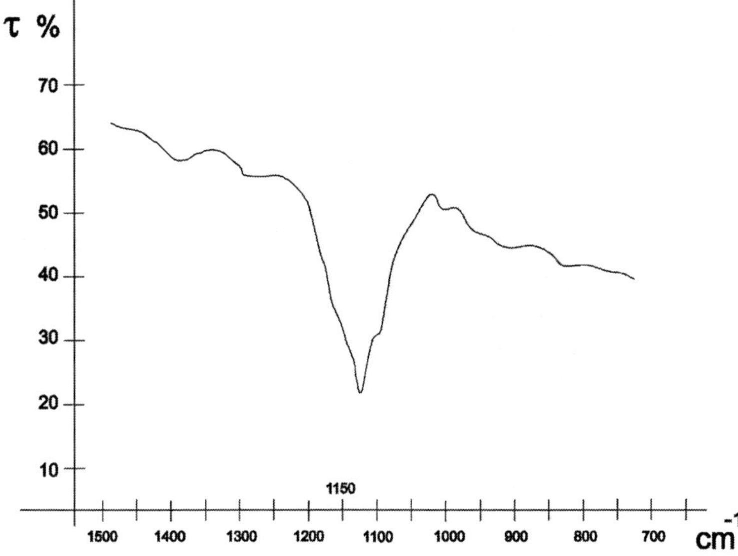

Figure 3. Infra Red (IR) Spectrum of Silicon-Dioxide Film.

Al-SiO_2 –Si-Al MIS structures were made and some electro-physical characteristics of the films were measured. Thickness and refractive index of films were measured by laser ellipsometer. Dielectric permeability (ε) and tangent of angle of dielectric loses (tgδ) were measured at 1KHz.frequency.

Properties of Dioxside Silicon Films				
thickness of film mcm	refractive index	dielectric permeability	dielectrid losses	electric strength V /cm
0,31-0,35	1,44	3,7	0,08-0,09	$E=10^6$

Table gives the data on the deposited layers. As is seen from the given characteristics, the deposited SiO_2 layers have good electro-physical properties and can be used both, in microelectronics and nanotechnologies.

CONCLUSION

The obtained results allow to use both, the pure SiO_2 layers and the multi-component insulation films deposited from the studied solution. On the basis of the obtained layers, it is possible to make nano-structurized materials using sol-gel technology. Besides, depending on the stated task, the practical solution of sol-gel method can be different.

REFERENCES

[1] M. Gotuaco et al. Implementation of CVD Low-K dielectrics For High Volume Production // *J. Solid State Technology*, 2004, N 1, p.29-37.

[2] T. Batchelder, W. Cai, J. Bremmer, D. Gray. In-Line Cure of SOD Low-K films // *J. Solid State Technology*, 1999, N3, p.29-34.

[3] N.P. Hacker. Organic and Inorganic Spin-on Polymers for Low-Dielectric-constant applications. // *J. MRS Bulletin*, 1997,N10, p. 33-38.

[4] V.A. Vasilev et al. Insulating Layers of Multi-layer Interconnection of Integrated circuits With Low Dielectric Permeability. "Elektronnaia promishlennost", 2004, p.145.

[5] G.A. Rasuvaev, B.R. Gribov, Г.A. Domrachev, B.A. Salamatin. Metal-organic compounds in electronics. "*Nauka*", Moscow, 1972, p. 479 (in Russion) .

[6] K.V. Zinoviev, O.F. Vikchliancev, B.G. Gribov. Production of Oxide Films From Solutions and Their Application in Electronic Engineering. 13(250), "*electronica*", Moscow, 1974, p. 127 (in Russion) .

[7] T.I. Pavliashvili, M.V.Janjalia, M.I. Gudavadze. Deposition of SiO_2 Films by the Method of Tetraethoxysilane Hydrolitic Decomposition. Georgia, *Chemical Journal*, 6(5), 2006, p.523-525. (in Russion).

In: New Developments in Material Science ISBN 978-1-61668-852-3
Editors: E. Chikoidze and T. Tchelidze

Chapter 19

INCREASE OF EFFICIENCY OF SOLAR CONCENTRATED ENERGY PHOTOELECTRIC TRANSFORMING MODULE

Gela Goderdzishvili, Alexander Moseshvili, Tengiz Mkheidze and Rafiel Chikovani *

Georgian Technical University,
77 Kostava, 0175 Tbilisi, Georgia

The last 10-15 years have been characterized by intensive research & develop on renewable energies including photovoltaic (PV) energy conversion. In this context the concentrator technology was researched more intensively. One reason is the availability of high-efficient A3B5 based triple-junction solar cells which have been originally developed for space applications. Applying now the concentrator technology an application on earth is possible. The significantly higher cost of the A3B5 technology can be compensated on system level and can be even lower compared to the standard silicon PV approach.

Eventually the lower system costs are achieved because of the high cell efficiency of about 40% and high resistance to highly concentrated solar radiation (more than 500x). Therefore, recently, the efforts of developers were directed to develop concentrator systems using A3B5 cell to enable commercially feasible terrestrial power generation with sufficiently low $/W.

* E-mail: rchikovani@mail.ru

The higher the concentration ratio and hence, the smaller area of solar cells, the lower the cost of the output energy.

In the indicated sphere the important achievements have the following Institutes and Companies: Ioffe Physico-technical Institute (Russian Academy of Sciences), Fraunhofer Institute – Solar Energiessysteme (Germany), Ferrara University (Italy), Sunlab (USA), ISFOC (Spain), Emcore Photovoltais (USA), Enea (Italy), Guascor (Spain), etc., the people working in them are the known specialists of this field: Dr. Viacheslav M. Andreev (Russia), Dr. Andreas W. Bett (Germany), Dr. Antonio Parretta (Italy), Dr. T.Zhang (U.S.A.), Gr. Guy R. Smekens (Belgium), Dr. M.Castro (Spain), etc.

The various concentrators for CPV modules are submitted in the following edition: Andreev V.M., Grilikhes V.A., Rumiantsev V.D., Leningrad "NAUKA" 1989, (RUS).

For today, the most widely used concentrator is the Fresnel lens. For the negative side of Fresnel lens can be deemed the complexity of execution, the necessity of precise aiming and the problems as a result of chromatic aberration which is a limiting condition for concentration degree. The enlisted problems negatively influence on parameters and price of CPV modules.

For overcoming the above indicated problems it has been designed a compound lens of a new type (Patent # P2008 4588B) and a concentrator of a new type (US patent) created through integrity of lenses and reflectors.

The compound lens of a new type is thin as a Fresnel lens but it doenot comprise the microscopic particles and is characterized by less chromatic aberration (Figure 1).

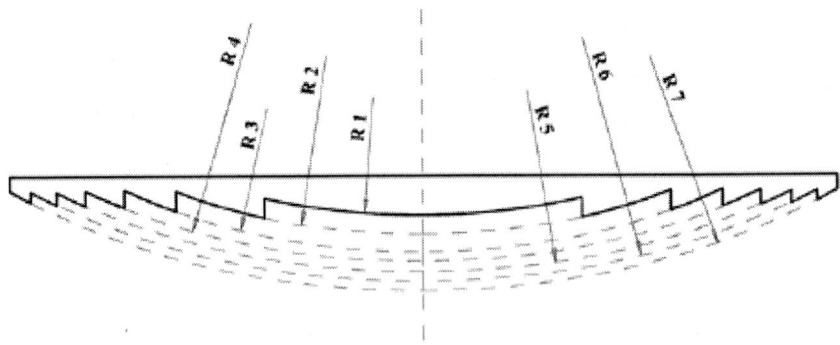

Figure 1. Cross section of compound lens.

It is comprised of several sections which considerably reduces the optical losses. Number of sections depends of such characteristics as the refraction index of lens material, focal distance and requirement to concentration degree.

By us it has been established that the small number of sections considerably reduces the losses caused by inaccurate aiming at the Sun. Each section is a part of spherical lens. The width of each section and the radius of surface curvature will be calculated so that each section will focus the beams incident on it in one and the same focal plane on one and the same area.

The method of execution of compound lens of a new type is simple, cheap and provides with achieving of high quality which has been proved by created testing lens (Figure 2).

The method of execution is new as well and maintaining of ideal sharpness of edges of sections is achieved without expensive programmable lathes. In parallel to creation of lens, collecting the solar radiation on the basis of semiconductor GaAs-AlGaAs heterostracture it has been processed both the construction of photoelectric transforming element and the method of its fabrication. Through using the methods of microelectronic technologies the assemblage of CPV experimental module has been realized.

Figure 2. Testing compound lens.

The new lens in comparison with a Fresnel lens, is characterized with a small chromatic aberration; Each unit of the new concentrator is completed with several small lenses with a long focal length and it is the large advantage, as chromatic aberration some more time decreases.

Much more reduction of chromatic aberration and height of the module is possible by means of combination of lens and reflectors. This construction is a novelty as well and is patented on International scale (A radiant energy concentrator - Patent # US 6,481,859 BI); In this new construction each semiconductor is served by several lenses, the focal spots of which are united by means of reflectors.

For modeling and optimization of a concentrator of a new type, a special computer program has been created; the interface and one computerized model of a concentrator has been shown on Figure 3.

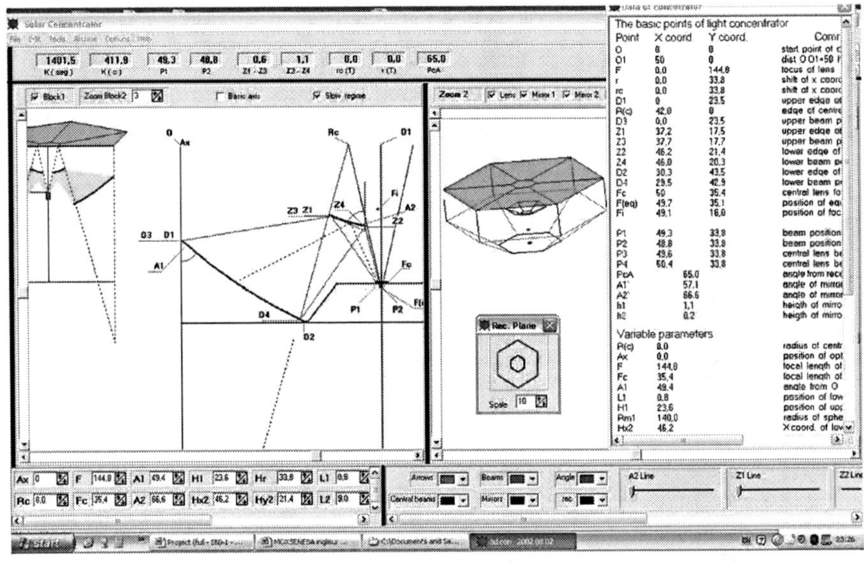

Figure 3. The interface and one computerized model of a concentrator.

Through the computer program is designed not only the construction of a concentrator as well is projected the distribution of density of beams on the semiconductor. In this particular model the further reduction of chromatic aberration, the desired magnitude is performed with concave spherical reflector. So as the chromatic aberration is progressively changed from the lens center, it will not managed to reduce to minimum the aberration of all beams by spherical reflector and it happens the optimization by means of special programming (Figure 4).

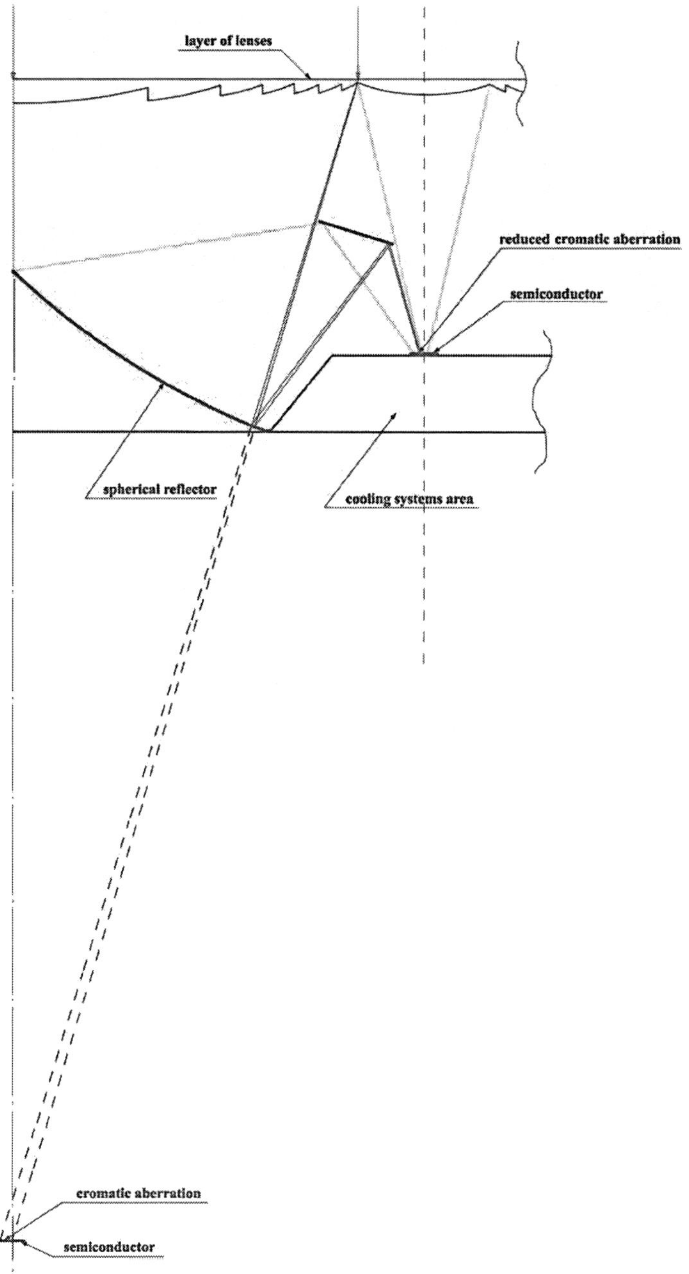

Figure 4. Process reduction chromatic aberration of one beam, from edge of a lens.

Figure 5. CPV module of one unit.

The working model created on the basis of computerized modeling (Figure 5) has been represented to International exhibition-conference in Barcelona.

This configuration has been processed constructionally and was created CPV module, completed with 42 units (Figure 6) which has been submitted to European Photovoltaic Solar Energy Conference and Exhibition in Milan. In this specific version of modeling, the semiconductor is located above from the basis of a module which gives a possibility to provide the cooling systems with liquid or gas by maintaining the same height of module (1 sq./dm. unit – 43.5 mm height).

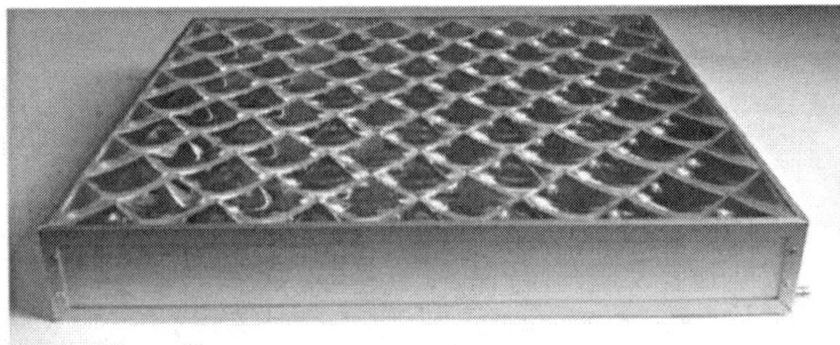

Figure 6. CPV module of 42 units.

To reduction of working areas of high-efficient, but expensive A3B 5 semiconductors impeded the limited possibilities of the concentrators available till present; The possibilities of represented concentrator (2000 and more), exceed the today's demands through maintaining the desired sizes. For the components of a concentrator there are selected the cheap and series production material and the method – is highly efficient (the main components are produced through integrated methods). According to preliminary price assessment of 1sq.m. of a concentrator should not exceed $50.

Patent includes additional novelty - multilevel concentrator (Figure 7) and with it is possible to achieve million and more concentration (9 streams of beams coming out from each previous level, get merged on each following level); heoretically at the third level it is possible to reach more then one million concentration. The purpose of multilevel concentrator is coming out of sphere of terrestrial CPV modules. Multilevel concentrator's peculiar properties (unlimited increase of concentration and a possibility of "unlimited" decrease of the height) make it incomparable for power supply in cosmos, for creation strong laser beams and in other spheres.

Figure 7. Cross section of multilevel concentrator (three level).

The represented work serves to make cheaper the energy worked out by the photovoltaic modules; As the concentrator of a new type gives a possibility to reduce much more working area of expensive semiconductors, the sizes and weight of the module and to raise energy-efficiency, is in the direct relation with downturn of price of ecologically pure and unexhausted energy and meets to International effort for increase of renewable energy share in the World's energy.

In: New Developments in Material Science ISBN 978-1-61668-852-3
Editors: E. Chikoidze and T. Tchelidze

Chapter 20

THERMALLY STIMULATED EVOLUTION OF STRUCTURAL DEFECTS IN LIF CRYSTALS IRRADIATED IN REACTOR AT DIFFERENT TEMPERATURES

***M. G. Abramishvili*[1], *M. V. Galustashvili*[1*], *G. G. Dekanozishvili*[1], *D. G. Driaev*[1], *T. L. Kalabegishvili*[1,2] *and V. G. Kvatchadze*[1]**

[1]E.Andronikashvili Institute of Physics, Tbilisi 0177, Georgia

[2]I.Chavchavadze State University, Tbilisi 0179, Georgia

ABSTRACT

The character of thermal recovery of mechanical properties (yield stress, plasticity, microhardness) of LiF crystals irradiated in the mixed (n, γ) field of reactor at 20 and 300 K temperatures has been studied. The evolution of radiation-induced microstructure is tracked, and it is shown that alongside with the disappearance of elemental defects and their complexes, the new structures – the prismatic dislocation loops of interstitial type and microcavities (pores) are formed.

---

* Corresponding author: maxsvet@yahoo.com

1. Introduction

Post-radiation high-temperature annealing of strongly irradiated crystal can finally cause a partial or even a full recovery of initial lattice and its properties, and thanks to this, the annealing is the main part of many technological processes. However, we think that for investigations, at present, more interesting and important, both from fundamental and applied points of view, are the intermediate stages of annealing, when, on the background of total destruction of radiation-induced formations, the new defect structures are formed that often lead to nontrivial results. Such "nontrivial course" of annealing became very valuable in the processes of creation of laser crystals [1], radiation detectors [1], crystal memory cells [2], etc. In this connection, we carried out a complex investigation of the results of post-radiation annealing of LiF crystals irradiated in the reactor channel at different temperatures.

2. Samples and Methods

LiF plates annealed preliminarily at 800°C were irradiated by a mixed (n, γ)-radiation at temperatures 300K and 20 K. The thermal neutron differential flux was 10^{12}n/cm^2·s and the fluence – 10^{16} and 10^{17}n/cm^2. The irradiated samples were subjected to isochronal annealing for 4 h at different temperature in the range of 200-800°C. Mechanical properties were studied by compression in the compression testing machine with the velocity of $2 \cdot 10^4$s^{-1}. Microhardness was measured by using Vickers pyramid. The pattern of sample etching was studied by metallographic microscope. The optical absorption spectra were measured in the 195-1100nm range. The curves of high-temperature thermally stimulated luminescence (TSL) were registered at the constant heating rate (1K/s).

Results and Discussion

Figure 1 presents the dependencies of yield stress τ and of maximum plastic deformation ε_m on the temperature of post-irradiation annealing of LiF crystals irradiated at 20K and 300K. It is seen that under the other similar conditions (irradiation dose, its rate, pre-radiation defect structure), the sample reveals a high radiation stability at low irradiation temperature (20K), as compared to the sample irradiated at 300K: the degree of hardening is less (half as much) and the plasticity is higher.

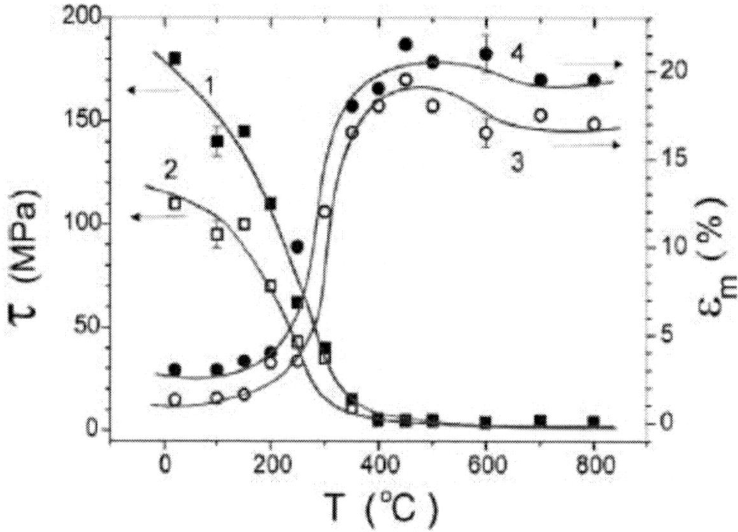

Figure 1. Dependence of yield stress τ (curves 1 and 2) and maximum elastic deformation ε_m (curves 3 and 4) on annealing temperature of LiF crystals irradiated (10^{16} n/cm^2) at 300 K (1 and 3) and 20K (2 and 4).

At post-irradiation annealing of crystals, the recovery of τ and ε_m takes place. At the temperature above 350°C, both of them reach the value almost equal to the initial (pre-irradiation) value: τ decreases and reaches the saturation, and ε_m passes through a small maximum and then remains, practically, constant.

Microhardness H behaves similarly as the yield limit τ, and the length l of rays of the dislocation rosette formed at indentation of crystal and characterizing the dislocation plasticity, recovers similar to the maximum deformation ε_m

Judging from the optical absorption spectra of crystals irradiated at 300 K and 20K, the radiation-induced structural defects revealed in them are qualitatively similar, but their quantity is more in the first case. At both irradiation temperatures the elemental color centers (CC) of F-type with the absorption maximum at 250 nm, and the aggregate centers of F_2- (450nm) and F_3-type (320 and 380 nm) are formed.

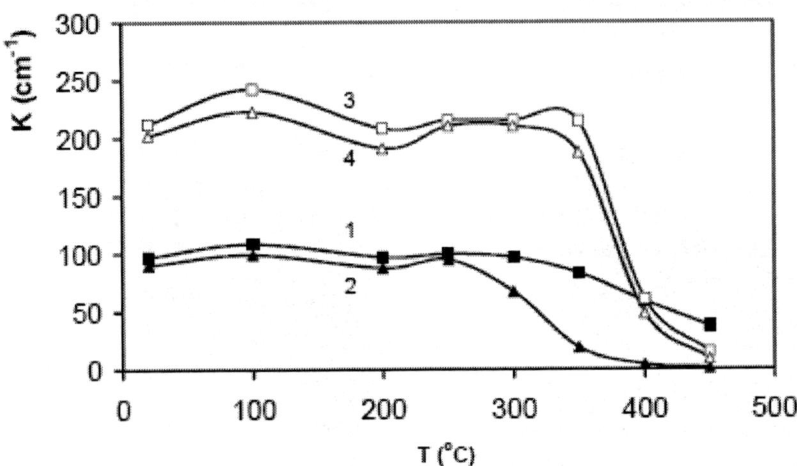

Figure 2. Dependence of optical absorption coefficient on annealing temperature of LiF crystals irradiated (10^{16} n/cm^2) at 20 K (1, 2) and at 300 K (3, 4). F-centers – curves 1 and 3, F_2-centers –2 and 4.

The dependence of optical absorption coefficients on annealing temperature (Figure 2) shows that at the early (first) stage of annealing (<250°C), the concentration of mentioned CC does not change significantly. Thus, the distinct recovery of mechanical characteristics taking place up to 250°C (Figure 1) cannot be connected with these structural defects. It is probably caused by the annealing of anisotropic radiation defects of bi-vacancy ($V_a^+V_c^-$) type, which are the strongest stoppers of dislocations and which significantly harden the irradiated crystal [3]. Having a rather low migration energy, these defects are annealed at relatively low temperatures, causing, thereby, a significant softening of the sample. And only above the temperature of 250°C (the second stage) the synchronism of the processes of CC annealing and the intense recovery of mechanical properties of the crystal are observed.

In contrast to F- and F_3-centers, the thermal stability of anisotropic F_2-centers depends on irradiation temperature (Figure 2). The active annealing of these centers in crystals irradiated at 20 K starts from 250°C (curve 2), and in the case of irradiation at 300 K – from 350°C (curve 4). This fact is also revealed in the measurements of thermally stimulated luminescence (Figure 3): the main TSL peaks of crystals irradiated at different temperatures are shifted relative to each other (along the temperature scale) by the same value as the points of starting the intense annealing of F_2-centers on curves 2 and 4 (Figure 2). Taking into account the similar literature data [4], it can be said that the observed peaks are mainly caused by these centers. The lower intensity of TSL in the case of annealing at 300 K is explained by a high density of radiation defects leading to the effect of luminescence quenching [5].

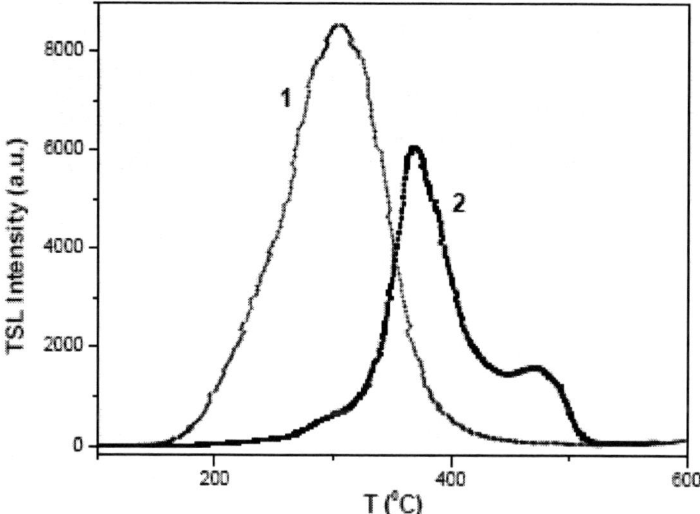

Figure 3. TSL curves of LiF crystals irradiated (1016 n/cm2) at 20 K (1) and at 300K (2).

A new peak with the maximum at 470°C is, probably, related to the dislocation electron zone [6], caused by dislocation loops observed at this annealing temperature. Besides, the value of this peak and the dislocation density increase with the irradiation dose.

Post-irradiation annealing, which, in its nature, should recover the initial crystal lattice, increases the difference between the crystals irradiated at different temperatures even more and, as it will be seen below, allows to observe visually this difference. First of all, this is expressed in formation of already mentioned dislocation loops, and then (at the increase of temperature) – of vacancy pores.

The pattern of etching of irradiated and annealed samples reveal the etching pits which are arranged in distinct pairs oriented along <100> and <110>, and which correspond to the prismatic dislocation loops of interstitial type [7]. They appear starting from 400°C and disappear above 500°C. The density of dislocation loops in the samples irradiated at 300 K is more than by an order of magnitude higher as compared to the density in the samples irradiated at 20 K and in both cases it increases with the irradiation fluence.

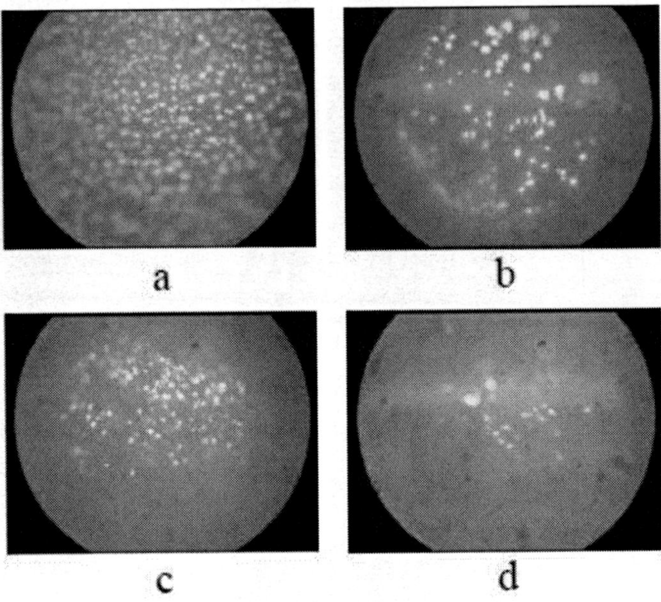

Figure 4. Pores on the fresh cleavage irradiated LiF crystal annealed at 700^{0}C. The fluence and the temperature of irradiation: a) 10^{17} n/cm^2 , 300K; b) 10^{17} n/cm^2, 20K; c) 10^{16} n/cm^2, 300 K; d)10^{16} n/cm^2, 20 K.

At the further heating of irradiated samples (≥600°C), the microcavities – pores are observed by the microscope (Figure 4). Pores have the shape of parallelepipeds oriented along <100> direction with the linear dimensions of 1-50 μm in two directions. The thickness of pore evaluated according to interference pattern is ≤0.4 μm. As it is seen from the Figure, the number and the dimension of pores depend on irradiation fluence and temperature, as well as on the temperature of post-radiation annealing. They are kept in crystals up to the melting temperatures. According to our preliminary data, there are the preconditions for creation of vacancy cells for hydrogenation of ion crystal by high-temperature annealing of strongly irradiated LiF crystals.

The solution of dislocation loops as well as the formation and the growth of pores do not have a remarkable influence on the yield limit and micro-hardness of the crystal. However, its plasticity decreases insignificantly with the increase of annealing temperature above 600°C, that is caused by the presence of pores being the potential site of crystal destruction [8].

CONCLUSION

For crystals irradiated in reactor at 20K and 300K, a similar spectrum of structural destructions is observed. However, in crystals irradiated at 300 K, the number of destructions are significantly more. Correspondingly, the degree of hardening is more and the plasticity is less.

It is shown that the thermal recovery of mechanical properties of irradiated samples above 250°C is caused by the annealing of CC. At early stages of annealing, the softening takes place at the expense of annealing of bi-vacancies.

Anisotropic F_2-centers reveal different thermal stability in crystals irradiated at 20 K and 300 K.

After the high-temperature annealing (>400°C) of irradiated crystals, we observe the formation of dislocation loops and later (at the increase of temperature to ≥600°C) – of vacancy pores kept in crystals up to the melting temperature. The number and the dimensions of dislocation loops and pores depend on temperature and irradiation fluence.

Dissolution of dislocation loops and the formation and the growth of pores have not a noticeable influence on the yield limit and micro-hardness of the crystal.

ACKNOWLEDGMENTS

The work has been fulfilled by financial support of the Georgia National Science Foundation (Grant #GNSF/ST08/4-414).

REFERENCES

[1] Nepomnyashchix, A. I.; Radjabov, E. A.; Egranov, A. V.; Color centers and luminescence of LiF crystals. *Nauka*, Novosibirsk. 1984.

[2] Shwartz, K. K.; Physics of optical recording in dielectrics and semiconductors. *Zinat*ne, Riga. 1986.

[3] Paperno, I. M.; Galustashvili, M. V.; *Fizika tverdogo tela*. 1976, 18(7), 1941-1943.

[4] Baldacchini, G.; Davidson, A. T.; Kalinov, V. S.; Kazakiewiez, A. G.; Montereali. R. M.; Nichelatti, E.; Voitovich, A. P.; *Phys. Stat. Sol.* (c). 2007, 4(3), 972-975.

[5] Las, W. C.; Stoebe, T. G.; Radiat. Protect. Dosimetry. 1984, 8, 45-67.

[6] Korovkin, E. V.; *Fizika tverdogo tela*. 2004, 46(6), 1013-1017.

[7] Andronikashvili, E. L.; Galustashvili, M. V.; Driaev, D. G.; Saralidze, Z. K.; *Fizika tverdogo tela*. 1987, 29(1), 130-135.

[8] Kukushkin, S. A.; Kuz'michev, S. V.; *Fizika tverdogo tela*. 2008, 50(8), 1390-1394.

INDEX

A

B

C

D

E

F

G

H

I

K

L

O

P

Q

R

S

T

U

V

W

Y

Z